BILLY HEATH

FOREWORD BY BRIAN POHANKA

The Man Who Survived
Custer's Last Stand

BILLY HEATH

VINCENT J. GENOVESE

Prometheus Books
59 John Glenn Drive
Amherst, New York 14228-2197

Published 2003 by Prometheus Books

Inquiries should be addressed to
Prometheus Books, 59 John Glenn Drive, Amherst, New York 14228–2197
VOICE: 716–691–0133, ext. 207; FAX: 716–564–2711
WWW.PROMETHEUSBOOKS.COM

07 06 05 04 03   5 4 3 2 1

Library of Congress Cataloging-in-Publication Data

Genovese, Vincent J., 1945–
    Billy Heath : the man who survived Custer's last stand / Vincent J. Genovese ; foreword by
Brian C. Pohanka.
        p. cm.
    Includes bibliographical references and index.
    ISBN 1–59102–066–2 (alk. paper)
    1. Heath, William, 1848–1891. 2. Little Bighorn, Battle of the, Mont., 1876. 3. United States.
Army. Cavalry, 7th—Biography. 4. Soldiers—United States—Biography. I. Title.

E83.876 .G47 2002
973.8'2—dc21
[B]   2002031912

Printed in Canada on acid-free paper

This book is dedicated to Schuylkill County, Pennsylvania, and its people, whose long and fascinating past now has yet another chapter to add to its colorful history: that of the Schuylkill County connection to the Battle of Little Bighorn.

CONTENTS

CONTENTS

ACKNOWLEDGMENTS

T he author is truly appreciative to the following people for their support in bringing William Heath's story to light: Rev. Bruce Sellers (pastor of Primitive Methodist Church, Tamaqua), William Gaydos (photographer), Dr. Richard Bindie (forensic pathologist), Donald Serfass (newspaper columnist), Ron Kensey (photographer), Jack Flynn (artist), Deborah Brumbaugh (great-granddaughter of William Heath), Richard Taylor (great-grandson of William Heath), Mark Major, Melanie Wade, and Geri Genovese (my wife). A special debt of gratitude is owed to John (Stu) Richards, a local historian and Custer scholar who unselfishly gave me a vast store of information on William Heath, and to Brian Pohanka, who reviewed the manuscript and provided me with sound and useful constructive criticism as did Dr. Louise Barnett. Thanks are also due to Dr. William Gudelunas for his afterword.

The help of the following institutions is recognized: Schuylkill County Court House Archives, *Pottsville Republican*, *Tamaqua Courier*, *Lehighton Times*, *Pottsville Miners Journal*, United States Army Military Records Archives, Tamaqua Primitive Methodist Church, Schuylkill

County Historical Society, Tamaqua Odd Fellows Cemetery, Minersville Public Library, Pennsylvania State University Library, Tamaqua Public Library, and Pottsville Public Library.

The author avers that the originality of this book is confined to the information directly related to William Heath. The historical descriptions in this work dealing with persons, places, and events other than William Heath are compilations of the various published works listed in the bibliography. In most cases, when exact wording was used in lesser-known statements it has been placed in quotes and the proper source identified. The more familiar remarks are simply in quotes and can be found in the listed works. In instances where information is presented in other than quotes the author is indebted to the creators of the listed works and does not take credit for them. There has been no conscious attempt by the writer to misrepresent any of these persons' works as that of his own. The happenstance of any such occurrence is completely unintentional, and, probably, literally unavoidable when writing on a subject wherein hundreds of books and articles have been published.

FOREWORD

The conflicts and controversies, theories and debates, drama and symbolism of the Battle of Little Bighorn have and will doubtless continue to attract the attention of researchers and buffs, journalists and historians. It is fertile ground for all manner of inquiry, and no battle of its size in all of history has spawned more text, art, and film than the story of "Custer's Last Stand."

Much of this interest is due to the larger-than-life persona of Lt. Col. and Bvt. Maj. Gen. George Armstrong Custer—a figure who inspires either adulation or scorn, and who overshadows and influences every aspect of that violent day. Custer's posthumous transformation to the realm of myth and symbol poses a challenge to all who seek to know the truth of his life, and of the factual circumstances of his death.

The complete destruction of the five companies under Custer's immediate command on June 25, 1876, is a further element in the ongoing fascination with Little Bighorn. Battles of annihilation, like large-scale disasters in the civilian world, have a perverse attraction for students of history—amateur and professional alike. "Last Stands" are by their nature tragic and heroic, controversial and symbolic—but they are also mysterious, for the simple fact that we can never know the precise sequence and details of what transpired among the fated and ultimately doomed command.

A recurring element in the literature of any such event are the questions, Was there a survivor? Did anyone escape the debacle? Did someone live to tell the tale? To delve into this question, to follow the cold trails and puzzle over the fragmentary evidence of such a possibility, is to risk entering the world of hoaxers, hucksters, and publicity seekers—those Little Bighorn historian William A. Graham aptly termed "the fabricators of fantastic tales about Custer's Last Fight." But amidst the hoaxes, delusions, and outright lies there still remains—at least in theory—a possibility that one (or more) of Custer's men did manage to escape from the fatal battleground.

There were, of course, a good many survivors of the battle of Little Bighorn, most obviously the several thousand Lakotas and Cheyennes who triumphed that day. And in fact more Seventh U.S. cavalrymen survived the engagement than perished in it. Despite losses, the companies comprising the battalions of Maj. Marcus Reno (A, G, and M) and Capt. Frederick Benteen (D, H, and K) escaped destruction, as did the soldiers assigned to accompany the mules of the regimental pack train: a detail from each company, some civilian packers, and an escort furnished by Capt. Thomas McDougall's Company B.

Those who perished on "Custer's Field" were the troopers of Companies C, E, F, I, and L—minus those detailed to the pack train—as well as a smattering of soldiers detailed to Custer's battalion from other companies, and several civilians. Most students of the fight accept the total of 210 as the number of men who died that day in the battalion under Custer's personal command. Contemporary accounts of the number of bodies interred at Custer's battleground vary from 202 to 212, with most estimates falling in the range of 204 to 208. Given the grim task at hand—the internment of the mutilated and decomposing dead—no one was immediately concerned with a precise enumeration of the fallen. Moreover, the condition of the slain often made individual identification impossible, though some effort was made to account for the officers.

Despite rumors and speculation that a dozen or more troopers had fought their way free of the debacle only to perish miles from the battlefield, the number of dead interred in 1876 is quite close to the number now believed to have accompanied Custer's battalion into the last fight. Over the years human remains have occasionally been found in the environs of Custer's Field, outside the present National Park boundary, while archaeological surveys within the park recovered numerous bones, and in one case a nearly complete skeleton. Whether some or all of these dead were counted in the initial tallies is open to conjecture. Postbattle eyewitnesses did document the discovery of three bodies between Custer's Field and the Reno/Benteen defense site: 1st Sgt. James Butler of Company L, Cpl. John Foley of Company C, and a trumpeter who most likely was Henry Dose of Company G, attached to Custer's headquarters detail. No one can say for certain if the three were felled in the opening stages of the fight, killed while bearing messages from Custer, or fleeing from the battle's desperate climax.

Initial confusion regarding the exact number of men who rode to their deaths with Custer's battalion stems in part from the fact that at least ten individuals who started northward with the five companies left the contingent prior to the Last Stand. Two of these were messengers dispatched by Custer: Sgt. Daniel Kanipe of Company C, and Trumpeter John Martin (Giovanni Martini) of Company H—the latter serving on Custer's headquarters detail. It was Martin who carried the famous note, scrawled by Regimental Adjutant William W. Cooke, urging Captain Benteen to "come on" and "be quick." In later years Pvt. Theodore Goldin of Company G also claimed to have been sent as a messenger, though his actual presence with the Custer battalion continues to be debated—and generally doubted—by students of the battle.

Four soldiers who started north with Custer's battalion were spared the fate of their comrades when they straggled behind the rapidly moving force. The lucky troopers—Pvts. John Brennan, John Fitzgerald, Peter Thompson, and James Watson—comprised a "set of four" in Company C. A veteran of the Seventh Cavalry later told researcher Walter Camp that it was the general opinion in the regiment that Brennan and Fitzgerald turned back "out of cowardice." In the case of Thompson and Watson evidence suggests that the exhaustion of their horses caused them to drop out of the battalion. The four ultimately rejoined Reno and Benteen and took part in the subsequent siege of the seven remaining companies. Some postbattle accounts suggest that a fifth trooper, Gustave Korn of Company I, may also have fallen back from Custer's battalion. A conclusive answer to Korn's whereabouts remains elusive, and it is possible that he was in fact assigned to the pack train.

Five Indian scouts accompanied Custer's northward-bound contingent, the acknowledged leader of whom, Mitch Boyer, perished with the battalion. The other four—White Man Runs Him, Hairy Moccasin, Goes Ahead, and Curley—departed the column prior to its annihilation. Just when and where they did so remains one of the most disputed subjects of a most disputatious event. Some believe Curley was the last to leave, and in fact made his escape somewhere in the vicinity of Calhoun Hill, when the final battle was already underway. Others note inconsistencies in Curley's later accounts and suspect that he left even earlier than his fellow scouts did. Whatever the actual facts of Curley's participation in the climactic fight, in his lifetime many considered the young Crow a "survivor" of Custer's last battle.

A little more than five weeks after the engagement at Little Bighorn soldiers of Gen. Alfred Terry's command (which included the survivors of the Seventh Cavalry) discovered a dead army horse on the Rosebud, near that stream's juncture with the Yellowstone River. The animal retained most of its equipment, and some witnesses recalled that a military carbine lay nearby. Although most veterans who recalled the event made no mention of a dead soldier, some alleged that a body, clad in cavalry attire, was found some miles upstream several days later. Whether fact, rumor, or a combination of the two, it was common talk within the ranks of the cavalry that the man in question was Pvt. Nathan Short of Company C; that either messenger or fugitive, he managed to escape from Custer's battalion only to perish on the banks of the Rosebud. The incident constitutes yet another mystery in the saga of Little Bighorn.

Over time dozens of tales of survivors of "The Custer Massacre" found their way into print; by 1952 author and bibliographer Fred Dustin had tallied "seventy-five or more." Many of them appeared in the press while legitimate veterans of Little Bighorn were still very much alive. In 1898 for instance, the *Army and Navy Journal* noted the story of George Benjamin, a crippled miner, who alleged that he was among the last seventeen troopers when the Indians overran Custer Hill. Taken alive "and subjected to horrible tortures" by his captors, Benjamin claimed to have been rescued three days after the battle by "Buffalo Bill and his cowboys." A somewhat more believable narrative emerged some four decades after the battle, when a respected farmer named Frank Finkel described his escape from the embattled ranks of Company C. Finkel's chronicle of his subsequent odyssey through the wilds of the West and ultimate salvation makes for interesting reading, but has generally been dismissed by serious students of the Little Bighorn.

Did farrier William Heath of Company L, Seventh U.S. Cavalry, survive the Battle of Little Bighorn? Perhaps the only valid answer to that question is, Let the evidence speak for itself.

Brian C. Pohanka
Alexandria, Virginia

PREFACE

The more serious reader might, at first glance, take note of the absence of footnotes in this work and erroneously assume it lacks any academic character. Such a presumption is far from the truth. Extensive research was done in a review of a representative sample of the vast body of literature on Little Bighorn. Although many published works are referred to and acknowledged, I chose not to footnote for two reasons.

First, my fear has always been that footnoting specific statements by individuals would bog the work down and hinder its readability. I wanted to avoid getting embroiled in the endless conflicting theories of what may have happened on that June day in 1876 and the criticism of which theories I chose to present. Arguments over troop movements, Custer's tactics, his motivations, and the like are infinite and would only detract from the book's main contention. It is not nearly so important where you are at any given point in a race as it is where you finish. Certainly, the final outcome of Custer's Last Stand is accurately conveyed.

A second reason for departing from a more academic approach was to not limit the reading audience to scholars only. My goal is to have a wide variety of readers evaluate this work and make up their own minds if its contention is, in fact, justified. It is for this reason that I refrain from waxing into long bouts of citing authorities and extemporizing on their views. This work is extensively documented in terms of William Heath's life and his movements throughout that life. It fully supports the likelihood that a least one man appears to have survived the famous Battle of Little Bighorn in which George Armstrong Custer and the men of his five companies were annihilated by Indians. The readers can decide for themselves if they agree.

A credible contention for a survival claim must have two key elements. First and foremost, one must establish that the person in question was present at the battle. The records from the archives of the United States Army confirm William Heath's presence at the battle and are reproduced in the exhibits. The second equally important criterion for the claim is to prove that Heath was alive after the struggle. The exhibits contain an abundance of documents that easily validate this. Thoughts such as mistaken identity or desertion are exercised in futility unless backed up by concrete evidence. Logical conclusions are only possible from the facts and the judgment reached is congruent with the known facts.

This work is somewhat critical of some of the Custer scholars. This reproach stems mostly from their refusal to consider the Heath story without ever reviewing the evidence. Despite this I have great respect for their academic credentials and I admire their passionate adherence to their theories. Perhaps this volume will prod them to reexamine their position.

One of this nation's most respected Custer scholars, Brian Pohanka, proved to be a great help to me during my research for this volume. Unlike many others, he did take the time to review the evidence I was able to compile, and he offered constructive criticism. Mr. Pohanka is well known for his books and lectures on the topic of Custer's Last Stand. He has been studying the Battle of Little Bighorn since the age of ten, visited the site over fifty times, and has taken part in photographic and archeological surveys of the battlefield. He may well be the consummate Custer scholar.

Throughout my research on Billy Heath, Pohanka's doubts were voiced as readily as his helpful suggestions. He corrected my drawing of the battlefield, questioned my view of General Custer, and pointed out that recent archeological studies are prompting scholars and Little Bighorn enthusiasts to reassess some of the long-held details regarding the battle. Like most of what has been written on this topic, these studies invoke new theories that as yet cannot be proven. It is fair to say that Brian Pohanka was intrigued by the Heath account but not convinced.

In one of our first exchanges Pohanka wrote, "I have learned in this topic that one must keep an open mind as there is so much we will never know. One must work to construct models of what might have happened. I have a very hard time imagining any trooper got away from the fight proper, but again, this is impossible to prove or dismiss out of hand." I could not agree more, and I believe I have constructed a model, based on facts, that points to the real possibility of the soldier named William Heath surviving Little Bighorn.

Oddly enough, Pohanka's next letter to me came after reviewing the manuscript while at Little Bighorn on the 125th anniversary of the battle. He was personally escorting British military around the battlefield. Here Pohanka said, after studying the struggle since the age of ten, he suspected he had become "in a sense too much involved in it." That same sensation has hit me after but several years of studying this infamous day in history. Pohanka commented: "Too bad Heath did not sit down and write his story as Frank Finkle did." Many a time I lamented the same lack of a written account. Unfortunately, Finkle "was not being truthful." Such is not the case here with the story of William Heath.

Brian Pohanka's main skepticism seems to center around a case of what he views to be mistaken identity. He writes: "So many men enlisted in the army under an alias, or assumed name, that somehow I think that is involved in this. . . . Many men who died at LBH cannot be located in any census records, leading me to think many of those used aliases." That is a fair criticism that deserves a response.

The theory of mistaken identity disregards some critical facts of the issue. The first of these is that an alias claim arbitrarily negates the evidence from an oral family history claim, which states that the Heath from Girardville, Pennsylvania, joined the army, served with the Seventh Cavalry at Little Bighorn, and returned home some months later. All of these facts are verifiable through existing government documents in the exhibits found in this volume, all of which leads one to the conclusion that the

Heath from Pennsylvania and the Heath acknowledged to have been out west were one and the same person.

Second, local tax records confirm William Heath's presence as a coal miner in Pennsylvania until the year of the battle when he is understandably absent. Following the battle his name was once again to be found on the tax rolls since he had returned home. If the Pennsylvania Heath and the one in the west were not the same, where was the Pennsylvania Heath in 1876 and who, in fact, was the Little Bighorn Heath? Both the timing of events coupled with the corresponding nature of the records make it highly unlikely that they were not the same.

Finally, the mathematical odds of the separate men matching up on so many details has to be prohibitive. Specifically, what are the odds that the unknown man joining the army picked the same name as a man from Pennsylvania whose family claims he, too, joined the army in the same year; served with the Seventh Cavalry; picked Staffordshire, England, as his birthplace; was the same age in years; and possessed the same physical characteristics (i.e., height, weight, hair color, etc.) as the Heath from Pennsylvania? Surely the calculations of such odds precludes coincidence.

Brian Pohanka and I have found some important common ground. One of my contentions supporting the survival claim is that the army made an understandable mistake in identifying William Heath as killed in action. Pohanka has written that due to mutilation, lack of clothing, no identification tags, and so forth, many soldiers were not identified. He states: "It was impossible. The U.S. Army was not interested in paying much attention to enlisted men in those days. In fact, there are only a relative handful of the enlisted men that were even tentatively ID'd from Co. L." These are simple statements with huge implications for the Heath mystery.

Another area of agreement focuses on the idea that the final jury with respect to Heath's survival should be the reader. Pohanka has said that the details of Heath's story "are somewhat tantalizing in that there is a hint of something there." He further comments of the tale that "like any such possibility it is deserving of investigation and consideration. The readers, the students, the historians, those drawn to that compelling episode of our history can choose to believe, disbelieve, or continue the investigation." I could not have said it better.

INTRODUCTION

Many Americans are passionate about the history of their country and its people. Harry S. Truman once remarked, "The only thing new in this world is the history that you don't know." British historian George R. Gissing wrote of history, "Persistent prophecy is a familiar way of assuring the event." Napoleon Bonaparte caustically commented that "History is a set of lies agreed upon."

The proliferation of books published on the history of this great nation boggles the mind. It numbers in the many thousands at the Library of Congress. No doubt it is a reflection of both our love for our country as well as our lifelong love affair with ourselves as a people. These narrations of past events come from every imaginable point of view. From the ultraconservative to the lavishly liberal they chronicle the happenings of our past. The many published accounts cover every facet of our 200-plus years of development, cultural diversity, military conflicts, ideas and beliefs, as well as our social and religious institutions. After all, diversity is often touted as our uniquely American trademark.

There is, however, one common thread to be found in all given accounts when the historical event under scrutiny is the famous massacre known variously as Custer's Last Stand or the Battle of Little Bighorn in

Montana. That common denominator is that every man under Gen. George Armstrong Custer's command of the five companies of the Seventh Cavalry died at that now infamous Battle of Little Bighorn. To emphasize this point I will cite from a handful of some of the best scholars ever to put a pen to American history, men whose speculations have guided us through the decades.

American History—The Modern Era Since 1865, by Donald A. Ritchie (2001), is one of the most widely used texts in the classrooms of today. Ritchie is an associate historian in the U.S. Senate Historical Office. Dr. Ritchie has taught U.S. history from high school to college and is a past president of the Oral History Association. In his book Dr. Ritchie states: "General Custer attacked, but the Native Americans killed Custer and all of his troops at the Battle of Little Bighorn."

H. E. Bourne and E. J. Benton wrote in their work, *A History of the United States in 1933*, "In a campaign against the Sioux in Montana, led by their Chief Sitting Bull, General George Custer, a young cavalry officer who had distinguished himself in the Civil War, and 264 of his troopers were suddenly surrounded and all of them killed." They later add, incorrectly, "only Custer's horse Comanche escaped."

An American History by Rebecca Brooks Gruver in 1972 recounts: "In June 1876 a large gathering of Sioux and their Cheyenne and Arapaho allies had encamped near the Little Bighorn River for the annual sundance ritual. One member of the three-prong military offensive moving towards them was a flamboyant and fiercely ambitious young Cavalry officer, Colonel George Armstrong Custer. . . . Custer split his seventh cavalry into three units and ordered an attack on the Indian camp. Not only did the soldiers fail to take the Indians by surprise, but they were vastly outnumbered. The two supporting wings were turned back, leaving Custer and some 264 men to face the majority of the camp's twenty-five hundred warriors. Although the troopers fought bravely, the entire battalion was wiped out."

The Story of America, by Herman M. Noyes (1964), details the events in these terms: "Custer was a daring cavalry officer. He graduated from West Point in 1861 and fought with outstanding bravery all through the War Between the States. In 1876 he was operating in what is now Montana. He divided his force in order to surround what he thought was a small group of Sioux Indians. Instead he ran into the great Chief Sitting Bull and a full war party. It was Custer's Last Stand. He and every man in his party were killed."

The distinguished *American History Illustrated*, which was published by the American Historical Society in 1971, ran an article written by Custer Scholar Robert Utley, who states in part, "For dramatic intensity, Custer's Last Stand can hardly be surpassed. He attacked a village that contained many more Indians than expected. The commander and every man of his immediate detachment fell in a death struggle with hordes of tribesmen." Utley adds, " 'only survivors' now outnumber Custer's command" and that "no survivor's story has withstood critical scrutiny."

The American Pageant, by T. A. Bailey (1961), is, in my opinion and that of many others, perhaps the best single book on American history ever written. The work reads like a novel as Bailey attempts (and succeeds) to show the actors in all the color and drama of the times, revealing their unique contribution to our history. Of Custer he writes: "The Sioux braves were hotly pursued by the lithe and impetuous George A. Custer, the buckskin-clad 'boy general.' Attacking what turned out to be a superior force of some 2,500 well-armed warriors near the Little Bighorn River in present Montana, the 'White Chief with Yellow Hair' and his 264 officers and men were wiped out in 1876 when two supporting columns failed to come to their rescue."

One of the most outstanding early accounts of the famous battle (published at the beginning of the last century) is the *Story of Little Bighorn* by Col. W. A. Graham, U.S. Army retired. Derived from primary sources still in existence, Graham wrote in his foreword: "The Battle of the Little Bighorn was a dramatic and romantic tragedy. It has provoked more speculation and more controversy than any other military episode in our history. It has a fascination for the military student that is unique. The wild and lonely country, the distance at the time from all settlements and all aid, the savage and fierce character of the Indian foe and his overwhelming numbers, as well as the great mystery which has forever shrouded the movements and actions of Custer's little command, and the manner in which he and five troops of the Seventh Cavalry met their deaths with not a single survivor, have excited the imagination and interests of soldiers and civilians alike for over sixty years."

Later, in the body of his work, Graham continues: "And so ended the

Battle of Little Bighorn. Upon the fatal field where Custer and his five companies fought and fell there were recovered more than 200 bodies. Some were never accounted for; but from that day to this, no trace of a single survivor has ever been found. They were utterly exterminated; not one escaped the fury of the Sioux." Of survivor claims of the times he states, "There is no truth to any of these tales. No authentic witness save the Sioux have ever appeared, and their accounts are at such variance that it is impossible to reconcile them."

Finally, there is the book titled *Son of the Morning Star* by Evan S. Connell (1984). It is one of the more distinctive current treatments of the battle. Connell's work is an unusual combination of part Custer biography, part Native American history. It gives insight into the human frailties and personalities of the primary players on the Little Bighorn stage. I include his work because, in addition to expressing belief in the total annihilation of Custer and his men, there are spurious references to the contrary. Connell observes that there "were more claimants than Custer had soldiers—of several hundred—who reputedly survived the Last Stand."

Despite his denial of the possibility of anyone escaping, Connell alludes to some evidence to the contrary. For example, "Several years after the battle two Cheyennes noticed a skeleton about 15 miles east of Custer Ridge—perhaps the remains of a trooper who had been stripped and left for dead but regained consciousness during the night and staggered away." In another interesting reference he tells us, "the Crow woman Pretty Shield said that for a long time her people used to find the bodies of soldiers and dead Indians far from the Bighorn." In the summer following the battle she said they discovered four dead blue coats together some six miles from the battle site. Connell's book mentions a person named Tom LeForge who often camped in the area of the battle and reported finding white men's remains some twenty-five miles from the Rosebud Valley (scene of the battle). Escapees? Perhaps, the historians tell us, but survivors are completely ruled out. But if remains could be found so far away from the battle is it not possible that some troopers survived?

Chapter One

EARLY YEARS
AND THE BLACK HOLE

onjecture is defined as an inference or an arrival at a conclusion based upon supposed facts which cannot be proven.
Anyone who has read the staggering body of literature on
Custer's Last Stand at the Battle of Little Bighorn can tell you that it is rife
with conjecture. There are probably more questions raised than answers
given when one studies this infamous battle; questions that (for lack of
clear evidence) will never be answered. Not that attempts haven't been
made. Theories offered by scholars, historians, and military strategists
abound, yet each falls short of a complete and thorough explanation of
the events on that fateful day. Historians of the battle differ on the exact
time it started, how long it lasted, troop movements, Custer's intentions,
number of fatalities, if Custer violated his orders, and if his tactics were
sound to name just a few.

Sweeping all the guesswork aside, facts that have recently been discovered reveal one irrefutable story that virtually destroys the long-
accepted belief that no one under Custer's command one hundred and
twenty-seven years ago survived the Battle of Little Bighorn. When the

dust kicked up by hundreds of horses settled on that blisteringly hot day in June, dead soldiers of the Seventh Cavalry were strewn everywhere. The stench of death permeated the battlefield when reinforcements arrived at the scene two days later. The United States Army grimly announced that Gen. George Armstrong Custer and every man in the five companies he commanded had perished. But startling new evidence reveals that one man, William (Billy) Heath, did, in fact, survive that horrible day. This is the true untold story of his life.

There are probably not more than a handful of Americans over the age of ten who are unfamiliar with Custer's Last Stand. The mystique of this story has not only endured but also grown over the last century and a quarter. Poking a rather large hole in a view that has been held so highly by so many for so long is sure to draw criticism. Nevertheless, for several reasons, I am convinced that this story must be told. First, there is the family of this survivor. For generations they have known the truth; the knowledge that at least one man did survive. Now that it is their desire for everyone to know what happened, I am eagerly willing to be their instrument.

Second, the saga of William Heath's life is one of the most compelling and intriguing series of events that I have ever come across. The manner in which this story unfolds and the elements of human interest it contains made it absolutely impossible for me to walk away. It cries out to be told. In my optimism I hope that, as you read on, you agree. Finally, and perhaps most importantly, is the matter of truth. I believe the time has come to embrace the facts and thereby correct the historical records of this important event. It is critical that our history books accurately impart the truth to all who read them. To fall short of anything but the truth would be doing a great disservice to the readers of this historical occurrence.

Although William Heath served with the Seventh Cavalry and lived most of his adult life in Schuylkill County, Pennsylvania, he was not born a citizen of the United States. Heath's life began on May 1, 1848, in Staffordshire, England. Two other years for his date of birth appear in public documents. One, an article in the *Lehighton Times* newspaper, lists his year of birth as 1846. A second, on his citizenship application in 1872 which lists his age at "23 or thereabouts," would make it out to be 1849. I believe the 1848 date to be the correct one for the main reason; not the least being that

I'm certain his wife, Margaret, made sure of the date when having his tombstone carved.

Staffordshire is a county in central England bounded by Cheshire on the north, Derbyshire on the east, Worcestershire on the south, and Shropshire to the west. It was here in Staffordshire at Tutbury and Chartley castles that Queen Elizabeth I held Mary Stuart captive. The county is rather diversified with pastureland for dairy farms and vegetables in the central area. Coal mines dominate the south. In addition, the county is known for its iron ore, marble, alabaster, clay, and fine pottery. Staffordshire is also the home of world-famous Wedgwood china.

Little is known of William Heath's earlier years in England before coming to the United States. Oral family history tells us that his occupation in England was that of coachman. The coachman was the taxicab driver or "hack" of the nineteenth century. He set down a footstool for passengers to use to enter the carriage, opened and closed the doors for them, and helped them alight from the cab. In addition, he took care of the horses and kept the carriage clean. His familiarity with horses would later play a pivotal role in his life. Nothing is known about Heath's father, whose name was either Sam or Simon, and his mother, Ann Baggally. William had a brother two years his senior named Arthur and a sister named Ruth three years his junior.

When William was yet a young child his family decided, like so many before and after them, to immigrate to the United States. No doubt, by all of the accounts of the times, the trip was long and hazardous. Immigrants of that era usually traveled on wooden three-masted vessels. The journey could take up to twelve weeks at the mercy of wind and weather. Below deck most of the time the conditions were very dangerous. The food supply frequently ran out or spoiled enroute. Sanitary conditions were deplorable and it was a given that not everyone who set sail at the outset would reach their destination alive. One can only imagine their relief when young William and his family safely reached the shores of America.

Filled with the hope of a promising life in a new country William's family settled in Schuylkill County, Pennsylvania. The county's land was purchased from the Delaware Indians in 1749. Originally these Indians were called Lenapes. They were not permanent residents of what is now Schuylkill County but returned there regularly to fish in the Delaware

River to the east and the Susquehanna on the west. The name Schuylkill is Dutch for "hidden river," which flows through the land.

The county was legally formed in 1811, carved from tracts of land that made up Berks and Lehigh Counties. By 1788 the indigenous tribes had already been successfully driven from the land. In 1780 Delaware tribesmen killed settler John Negman and his three children at the present site of the city of Pottsville. This caused the few settlers in the area at the time to leave, but a short time later they returned.

At the southeast end of the county is a part of the Appalachian Mountain range known as the Blue Mountains. The highest of these is 1,700 feet above sea level. The Schuylkill River, which flows unimpeded to Philadelphia, drains the county's 767 square miles. Until the advent of railroads the river served as the main means of travel to the city of Brotherly Love and was the lifeline of commerce to the residents of Schuylkill County.

The discovery and development of commercially useful hard anthracite coal was the key to the economic growth of the county. It was in the mines that William Heath would eventually go to work trying to make a living for himself and his family. Understanding the conditions that existed in Schuylkill County as Heath reached young manhood is key to grasping the circumstances that led him to the far-removed battlefield of Little Bighorn.

The discovery of coal in Schuylkill County is usually credited to Necho Allen. In 1790 Allen was a lumberman living somewhat of a vagrants' life seeking out the virgin tree stands in parts of the country. One day, while searching the Broad Mountain near Pottsville, nightfall came and he decided to camp for the night. After building his fire he fell asleep. Allen woke up in the middle of the night to the feeling of intense heat coming from a mass of glowing rocks. Apparently he had accidentally ignited an outcropping of a pure coal vein. Seeing this mass of glowing fire, Necho Allen became aware of its possibilities. When he returned to town he vigorously advocated its use. Sadly, his proddings fell on deaf ears. Allen left the county in disgust and never returned.

By 1795 there are reports that some of the blacksmiths in the area were using the black rocks. In 1812 was recorded the first attempt to mass market the coal of the country. Col. George Shoemaker, the proprietor of the Pennsylvania Hall Hotel in Pottsville, loaded up nine wagons with the black rocks and transported them to Philadelphia. Initially, the residents of the city denounced him as a swindler and an impostor. However, once he demonstrated the rock's heat-producing capabilities the wagons were quickly sold out and demand for coal grew.

Donald L. Miller and Richard E. Sharpless report in their book *The Kingdom of Coal* that in 1826 Abraham Potts, the founding father of Pottsville, invented a new system for conveying coal from his Black Valley mine in Schuylkill County. News of the contrivance soon reached the city of Philadelphia and a number of officers and gentlemen of means made the trip to see for themselves. Upon their arrival they found thirteen fully loaded coal cars seated atop wooden rails that ran from the mine opening to a navigation canal at a place called Mill Creek. The ever-agreeable Abe quickly offered a demonstration. He hitched the cars up to a single horse (amid much wisecracking from the observers from the city) and had the animal easily pull the load to its destination along the canal. Potts boasted that within a decade coal would be moving all the way to Philadelphia by rail. The Philadelphians left unimpressed, vowing to lock Potts up for lunacy if he ever set foot in their city.

Nevertheless, coal began moving to market by way of Philadelphia. In 1820 365,000 tons were shipped to the city. Most of it traveled by a series of canals and the Schuylkill River. As part of the canal system a tunnel was driven through the mountain at Port Carbon to transport the barges. It was an engineering marvel and the first one in North America. Taking three years to complete, the tunnel was 460 feet long. Men and women of leisure came by stagecoach from as far away as Philadelphia, 100 miles removed, to witness the spectacle.

By 1847 the number of tons shipped grew nearly tenfold to three million. Potts's prediction had finally materialized: most of the coal now traveled by rail. On New Year's Day in 1842 the first locomotive made the trip from Philadelphia to Schuylkill County. In 1872, in the post–Civil War era, no less than nineteen million tons reached Philadelphia and almost 11,500 ships cleared the city's port with coal bound for neighboring states and foreign countries.

The Civil War alone dramatically increased the demand for coal. Following the war the Industrial Revolution and the skyrocketing expansion of the country led to booming prosperity for Schuylkill County and its coal mines. The popular appeal of coal along with the many uses to which it could be put (especially in the steel and iron industries) fanned the flames of growth. Mines opened everywhere. Roads and train tracks were built over and through the mountains, with towns springing up like mushrooms. Small tracts of land bought for five hundred dollars in 1827 sold for sixteen thousand dollars two years later. The high demand for the black gold drove up its price as well as the miners' wages. Coal was now accepted as a proven fuel worldwide, assuring the county's long-term economic prosperity.

That was the good news. But as Billy Heath was approaching his mid-twenties there were disturbing signs on the horizon regarding the progress Schuylkill County had enjoyed. The mid-1870s ushered in, to everyone's surprise, a harsh nation-wide recession due to overspeculation in the stock market and a lack of an adequate supply of government gold reserves. The coal regions, well noted for their aggressive labor movements, changed into places of turmoil. Management, equally combative, butted heads with the miners. Strikes ensued. Unemployment, violence, and mayhem were the order of the day in Girardville, Pennsylvania, where Heath was living.

The little town of Girardville was founded by one of the nation's wealthiest men, philanthropist Steven Girard of Philadelphia. He wisely purchased a large tract of land in Schuylkill County. The town that was named after him sprung up over night with the discovery of coal nearby. When the Civil War began, Girardville probably had fewer than one hundred residents. By 1875 it claimed three thousand permanent citizens. Within several miles of the town there were at least ten collieries pushing out the finished coal from scores of mines in the area. It was here in the booming coal town of Girardville that William came to live when he first arrived in Schuylkill County.

On September 30, 1872, a more documented image of William Heath begins to appear, for on that date he officially applied for citizenship in the United States of America (see exhibit 1). As can be seen from the petition, Heath was between twenty-three and twenty-four years old, having been in this country at least three years prior to the application. His sponsor was his father-in-law, Joseph Swansborough. William had married Margaret Swansborough just a few days before, on September 21, 1872. It is entirely possible that Joseph Swansborough insisted that his son-in-law become a U.S. citizen. Being an influential man in the politics of the Town of

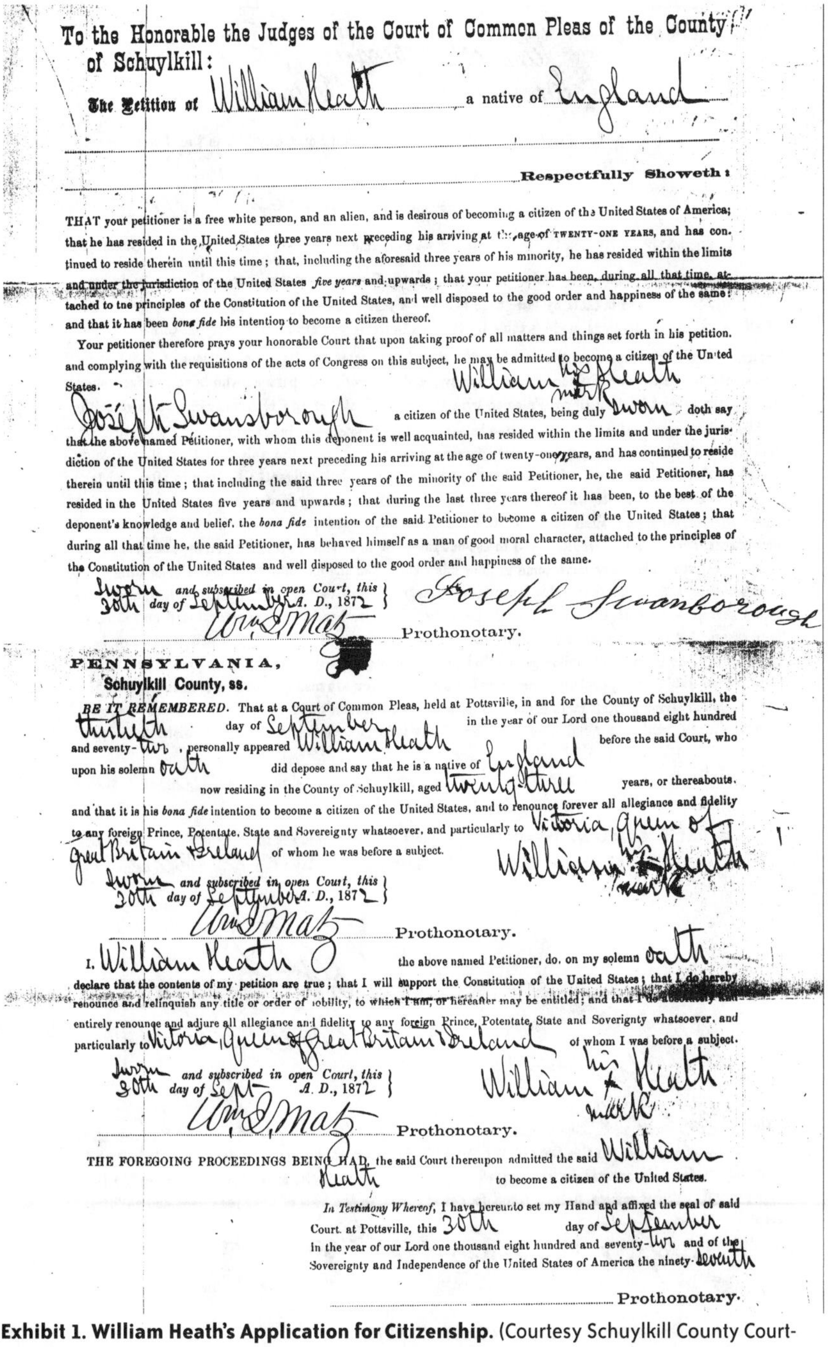

To the Honorable the Judges of the Court of Common Pleas of the County of Schuylkill:

The Petition of William Heath a native of England

Respectfully Showeth:

THAT your petitioner is a free white person, and an alien, and is desirous of becoming a citizen of the United States of America; that he has resided in the United States three years next preceding his arriving at the age of TWENTY-ONE YEARS, and has continued to reside therein until this time; that, including the aforesaid three years of his minority, he has resided within the limits and under the jurisdiction of the United States *five years* and upwards; that your petitioner has been, during all that time, attached to the principles of the Constitution of the United States, and well disposed to the good order and happiness of the same: and that it has been *bona fide* his intention to become a citizen thereof.

Your petitioner therefore prays your honorable Court that upon taking proof of all matters and things set forth in his petition, and complying with the requisitions of the acts of Congress on this subject, he may be admitted to become a citizen of the United States.

William his mark Heath

Joseph Swansborough a citizen of the United States, being duly sworn doth say that the above named Petitioner, with whom this deponent is well acquainted, has resided within the limits and under the jurisdiction of the United States for three years next preceding his arriving at the age of twenty-one years, and has continued to reside therein until this time; that including the said three years of the minority of the said Petitioner, he, the said Petitioner, has resided in the United States five years and upwards; that during the last three years thereof it has been, to the best of the deponent's knowledge and belief, the *bona fide* intention of the said Petitioner to become a citizen of the United States; that during all that time he, the said Petitioner, has behaved himself as a man of good moral character, attached to the principles of the Constitution of the United States and well disposed to the good order and happiness of the same.

Sworn *and subscribed in open Court, this* 30th *day of* September *A. D.,* 1872

Joseph Swanborough

Wm. D. Matz Prothonotary.

PENNSYLVANIA,
Schuylkill County, ss.

BE IT REMEMBERED. That at a Court of Common Pleas, held at Pottsville, in and for the County of Schuylkill, the thirtieth day of September in the year of our Lord one thousand eight hundred and seventy-two, personally appeared William Heath before the said Court, who upon his solemn oath did depose and say that he is a native of England now residing in the County of Schuylkill, aged twenty three years, or thereabouts, and that it is his *bona fide* intention to become a citizen of the United States, and to renounce forever all allegiance and fidelity to any foreign Prince, Potentate, State and Sovereignty whatsoever, and particularly to Victoria, Queen of Great Britain & Ireland of whom he was before a subject.

William his mark Heath

Sworn *and subscribed in open Court, this* 30th *day of* September *A. D.,* 1872

Wm. D. Matz Prothonotary.

I, William Heath the above named Petitioner, do, on my solemn oath declare that the contents of my petition are true; that I will support the Constitution of the United States; that I do hereby renounce and relinquish any title or order of nobility, to which I am, or hereafter may be entitled; and that I do entirely renounce and adjure all allegiance and fidelity to any foreign Prince, Potentate, State and Soverignty whatsoever, and particularly to Victoria, Queen of Great Britain & Ireland of whom I was before a subject.

Sworn *and subscribed in open Court, this* 30th *day of* Sept *A. D.,* 1872

William his mark Heath

Wm. D. Matz Prothonotary.

THE FOREGOING PROCEEDINGS BEING HAD, the said Court thereupon admitted the said William Heath to become a citizen of the United States.

In Testimony Whereof, I have hereunto set my Hand and affixed the seal of said Court, at Pottsville, this 30th day of September in the year of our Lord one thousand eight hundred and seventy-two and of the Sovereignty and Independence of the United States of America the ninety-seventh

Prothonotary.

Exhibit 1. William Heath's Application for Citizenship. (Courtesy Schuylkill County Courthouse Archives)

Schuylkill County, ss. *Patrick Dormer, Valentine Benner and Moses Hine*

Esquires, Commissioners of the County of Schuylkill, in the Commonwealth of Pennsylvania, To *Robert Green*

Assessor of the *Borough of Girardville*

in the County and State aforesaid, - - - - - GREETING :

YOU are hereby required to take an account of all *Freemen,* and the personal property in your *Boro* made taxable by law; together with a just valuation of the same, and also a valuation of all trades and occupations made taxable by law; and of this you are hereby required to make a just return, within thirty days from the date hereof, noting in the same all alterations in your *Boro* occasioned by transfer or division of real property; and also noting all persons who have removed since the last assessment, and all single Freemen who have arrived at the age of twenty one years since the last triennial assessment, and all others who have since that time come to inhabit in your *Boro* together with the taxable property such persons may possess, and the valuation thereof, agreeably to the provisions of the law. You are hereby further commanded, diligently to inquire after and take an account of all real estate which since the first day of May last may have passed from persons dying seized thereof, otherwise than to or for the use of father, mother, husband, wife, children and lineal descendants, born in lawful wedlock, and to set out the same in schedule attached hereto; and also of any estate or estates which may have come to your knowledge from any executor, administrator or otherwise, together with a fair and just valuation on the same, according to the market price thereof, and make return thereof to us with the list of taxable property. You are hereby further commanded and directed to re-assess, between the periods of the triennial assessments, all real estate which may have been improved by the erection of buildings or other improvements, subsequent to the last preceding triennial assessment—subject to appeals, as now provided by law. You are also required to furnish a list to the County Commissioners of all male persons in your district between the ages of 21 and 45 years, subject to militia duty, with the number of taxables of any independent School district or districts in your

THE DAY OF APPEAL will be held at the *Commissioners Office* on the *any* day of *May* next.

In Witness Whereof, the said Commissioners have hereunto set their hands and affixed the seal of the said Office, at Pottsville, this *first* day of *April* A. D. 18*73*.

J. Benner

Moses Hine } Commissioners.

Attest: *O. J. Alsgood* Clerk.

You, *Robert Green* elected Assessor for the *Borough of Girardville* in the County of Schuylkill, do swear, that you will diligently, faithfully and impartially perform the several duties enjoined on you as Assessor of the said *Borough* by the above precept, without malice, hatred, favor, fear or affection, to the best of your judgment and abilities. Sworn and subscribed before me, this *19* day of *April* A. D. 18*73*.

Henry Shopstall

Justice of the peace

Exhibit 2. Cover Page. Girardville Tax Records 1873.
(Courtesy Schuylkill County Courthouse Archives)

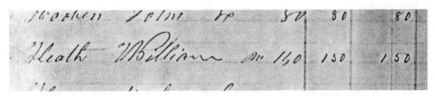

Exhibit 2. Girardville Tax Records 1873. (Courtesy Schuylkill County Courthouse Archives)

Girardville, Swansborough could ill afford a son-in-law who was not an American. It is very likely that it was Swansborough who secured William his first job working in the mines. Mr. Swansborough himself was listed as a miner by occupation and may have been looking out for the well-being of his daughter by making sure her husband had a decent job so he could provide for his family.

This document is important as it establishes the existence of the man named William Heath living in Schuylkill County, roughly twenty-four years old, and originally coming from England. This information corroborates the oral family history and each gives the other credibility.

The Heaths set up housekeeping in Girardville and William found work digging coal in the mines. The tax records from the county archives list William Heath as a miner living in Girardville and paying a yearly assessment of fifteen dollars for the years 1873 and 1874. During this time Margaret and William were excited newlyweds but otherwise theirs was an uneventful life in the scheme of what was about to happen. It is also worth noting that Joseph Swansborough served as Chief Burgess (mayor) of Girardville in the years 1873 and 1874.

Exhibit 2 shows William Heath's property listed on Girardville's tax records for 1873 at an assessed value of one hundred fifty dollars. Exhibit 3 shows Joseph Swansborough living in Girardville with the same assessment. Both men are listed as miners. In 1874 Heath's assessment has more than doubled to four hundred dollars, as shown on exhibit 4. He was enjoying a much better year financially. In 1875, perhaps as a result of the

Exhibit 3. Girardville Tax Records 1873. (Courtesy Schuylkill County Courthouse Archives)

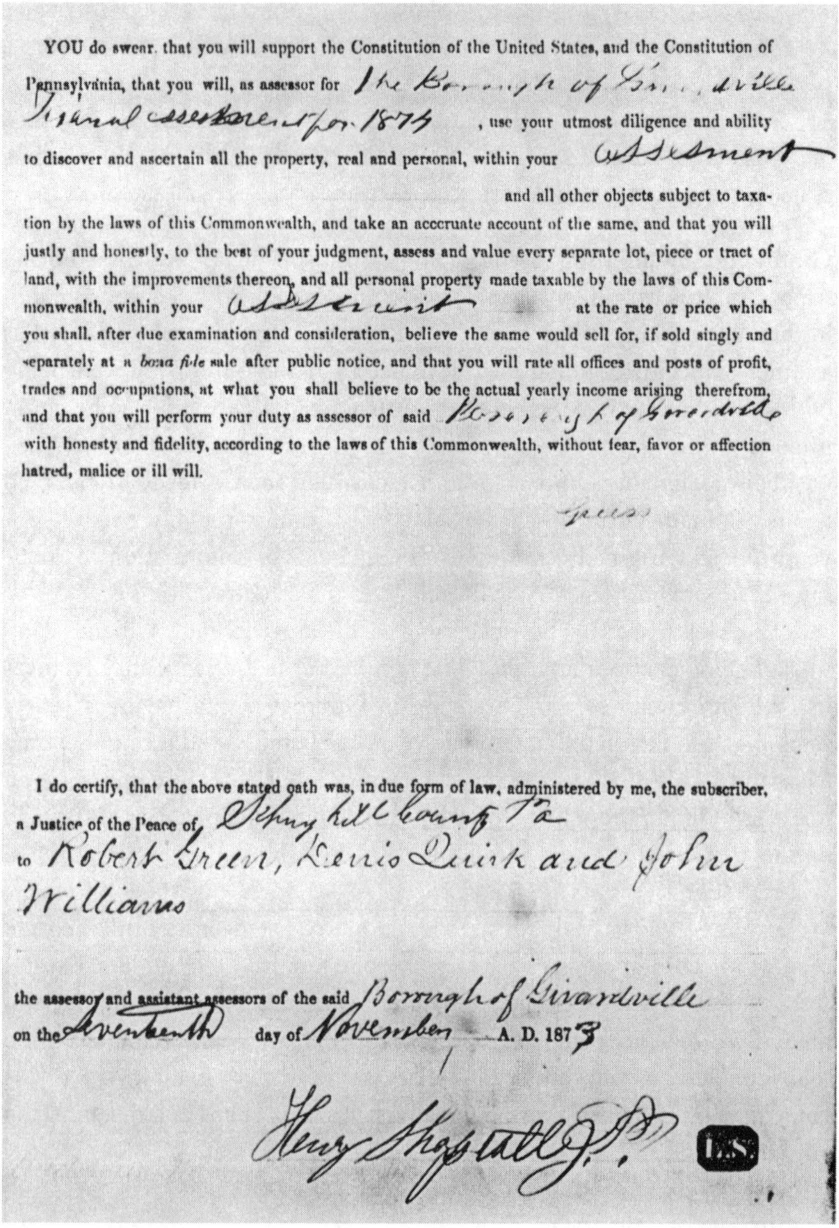

YOU do swear, that you will support the Constitution of the United States, and the Constitution of Pennsylvania, that you will, as assessor for *The Borough of Girardville* *assessment for 1874* , use your utmost diligence and ability to discover and ascertain all the property, real and personal, within your *assessment* and all other objects subject to taxation by the laws of this Commonwealth, and take an accruate account of the same, and that you will justly and honestly, to the best of your judgment, assess and value every separate lot, piece or tract of land, with the improvements thereon, and all personal property made taxable by the laws of this Commonwealth, within your *assessment* at the rate or price which you shall, after due examination and consideration, believe the same would sell for, if sold singly and separately at a *bona fide* sale after public notice, and that you will rate all offices and posts of profit, trades and occupations, at what you shall believe to be the actual yearly income arising therefrom, and that you will perform your duty as assessor of said *Borough of Girardville* with honesty and fidelity, according to the laws of this Commonwealth, without fear, favor or affection hatred, malice or ill will.

I do certify, that the above stated oath was, in due form of law, administered by me, the subscriber, a Justice of the Peace of *Schuylkill County Pa*

to *Robert Green, Lewis Quirk and John Williams*

the assessor and assistant assessors of the said *Borough of Girardville* on the *seventeenth* day of *November* A. D. 187*3*

Henry Shappell J.P.

Exhibit 4. Cover Page. Girardville Tax Records 1873.
(Courtesy Schuylkill County Courthouse Archives)

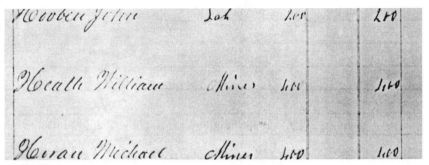

Exhibit 4. Girardville Tax Records 1874. (Courtesy Schuylkill County Courthouse Archives)

labor unrest discussed later in this chapter, his property assessment is down 25 percent to three hundred dollars (see exhibit 5). These exhibits confirm that Heath was living in Girardville and working in the coal mines as family history relates. His paying taxes and keeping a residence is further proof that he was alive and well in the years reported.

The significance of the development of the coal industry in Schuylkill County is important to the Heath story for several reasons. Working in the mines seems to have been a natural connection to his past. Family history relates that William's father worked in the coal mines in Staffordshire, England. When the family came to this country, they sought a place where work in a familiar setting could be found. William's first job of record was that of a laborer in the mines of Girardville. It is where he toiled before the Battle of the Little Bighorn and whence he returned afterward.

Heath fed himself and his family by working in the mines. It is what his father-in-law's occupation was. Several of Heath's sons earned their livelihood in the mines after Heath's death. The entire existence of the family was centered on the nearby mines, the coal-patch town they lived in, the company store that provided their daily staples, and the social life that the subsistence of a coal miner offered. Although Heath may go down in history for his exploits as a soldier with the Seventh Cavalry, most of his adult life was spent deep underground working the black hole.

Lastly, the economic developments that took place in the coalfields of the county circa 1875 directly relate to how William Heath ended up in Montana in June of 1876. The dramatic upheaval that took place in Schuylkill County changed Heath's life forever. So dramatic were these events that he left behind his job, his home, and his wife and child. One can

SCHUYLKILL COUNTY, *ss.* *Moses Hine, Patrick Coury and Morgan W. Fehr*

Esquires, Commissioners of the County of Schuylkill, in the Commonwealth of Pennsylvania, To *Thomas Fletcher* Assessor of the *Boro of Girardville*

in the County and State aforesaid, ... GREETING:

YOU are hereby required to take an account of all Freemen, and the personal property in your *Boro* made taxable by law; together with a just valuation of the same, and also a valuation of all trades and occupations made taxable by law; and of this you are hereby required to make a just return, within thirty days from the date hereof, noting in the same all alterations in your *Boro* occasioned by transfer or division of real property; and also noting all persons who have removed since the last assessment, and all single Freemen who have arrived at the age of twenty-one years since the last triennial assessment, and all others who have since that time come to inhabit in your *Boro* together with the taxable property such persons may possess, and the valuation thereof, agreeably to the provisions of the law. You are hereby further commanded, diligently to inquire after and take an account of all real estate which since the first of May last may have passed from persons dying seized thereof, otherwise than to or for the use of father, mother, husband, wife, children and lineal descendants, born in lawful wedlock, and to set out the same in schedule attached hereto; and also of any estate or estates which may have come to your knowledge from any executor, administrator or otherwise, together with a fair and just valuation on the same, according to the market price thereof, and make return thereof to us with the list of taxable property. You are hereby further commanded and directed to re-assess, between the periods of the triennial assessments, all real estate which may have been improved by the erection of buildings or other improvements, subsequent to the last preceding triennial assessment—subject to appeals, as now provided by law. You are also required to furnish a list to the County Commissioners of all male persons in your district between the ages of 21 and 45 years, subject to militia duty, with the number of taxables of any independent School district or districts in your *Boro* .

THE DAY OF APPEAL will be held at the *Commissioners Office* on the *18th* day of *May* next.

IN WITNESS WHEREOF, the said Commissioners have hereunto set their hands and affixed the seal of the said office, at Pottsville, this *first* day of *April* A. D. 187*5* .

Moses Hine

Patrick Coury

M. W. Fehr

Commissioners

Attest: *O. J. Augord* Clerk.

YOU, *Thos Fletcher* elected Assessor for the *Borough of Girardville* in the County of Schuylkill, do swear, that you will diligently, faithfully and impartially perform the several duties enjoined on you as Assessor of the said *Borough* by the above precept, without malice, hatred, favor, fear or affection, to the best of your judgment and abilities.

Sworn and subscribed before me,

Exhibit 5. Cover page. (Courtesy Schuylkill County Courthouse Archives)

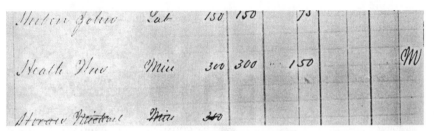

Exhibit 5. Girardville Tax Records 1875. (Courtesy Schuylkill County Courthouse Archives)

only wonder how a man could make such a bold move and whether or not the reasons for making it were that clear and present a danger. Obviously to William Heath they were.

Unlike William Heath, most men began working in the mines at an early age (some by the time they were six years old, but certainly by the time they were eight or nine).* They started their careers in the breakers and were known as "breaker boys." The breaker was a pyramid-shaped building that rose some 150 feet into the air, dwarfing all other buildings around it. The odd shape of the building was designed to enable form to serve function. After the coal was dug it was hauled to the top of the structure in cars and tipped over. The raw coal was crushed by revolving cylinders and screened into various sizes. Gravity forced the coal down along a network of chutes. The young breaker boys straddled these chutes with feet dangling to slow the coal's progress. Here they picked by hand the slate and other refuse mixed with the coal so the end of the run produced only pure coal (see exhibit 6).

Nearly one in four mine workers was one of these lads. In the predawn darkness fathers carried their still-sleeping sons on their backs to the breakers to work above ground as they descended the shafts. Often mothers would arrive at lunchtime with a dinner pail since many of the lads were too young to handle one by themselves. Before the days of using water the dust would be so thick they couldn't see each other on the chutes. They wore rags over their noses and chewed tobacco to keep from choking on the black dust.

These youngsters toiled at the breakers for ten hours a day six days a week for forty-five cents a day. There were no gloves allowed on the job and in the winter, when snow covered the ground, the path of the young

*Much of the information on mining and Schuylkill County is taken from *The Kingdom of Coal*, by Donald L. Miller and Richard E. Sharpless, and *The Reading Railroad*, by James Holton.

Exhibit 6. Breaker. (Courtesy Don Serfass)

boys' homeward trek could be traced by the drips of blood from their hands (see exhibit 7).

A company foreman was constantly on hand to maintain order and keep everyone working. The boys, who were prone to daydreaming, falling asleep, and gabbing with one another, were kept in line with harsh measures. Clubs and leather switches were used in what was known as "whipping them in." Under these conditions the youths quickly lost their naiveté and became outspoken and rebellious in spite of the consequences. These little wards developed a hatred for all bosses, which would later mature into outright violence when they went to work as full-fledged miners deep in the bowels of the earth.

At the end of their shift they ran like ants from their prison, happy to be free at last until the next day's work. Home they went with faces and teeth blacked by coal dust to get a wash and some supper. Afterward, if they weren't too tired, some would go out to play while others might attend night school. Those who had a studious nature could inevitably wear down the hardiest of teachers who were noted for not lasting long in the unruly schoolroom of a coal-mining town. Surely these children were among the rowdiest and most aggressive children in the entire nation. It was a scene novelist Stephen Crane depicted with horror in one of his works after paying a visit to a breaker in Schuylkill County.

The world that William Heath descended into each morning was one almost beyond imagination. It is very often described as working in a black hell. Before the crack of dawn

Exhibit 7. Miners walking to breaker. (Courtesy Don Serfass)

Exhibit 8. Coal miner. (Courtesy artist Jack Flynn)

Heath would be awakened by the blast of the colliery whistle. He would walk to work dressed in his black-encrusted clothes and boots, wearing a miner's lamp on his hat for light and carrying his lunch pail (see exhibit 8). Young William was transported to his work site hundreds of feet below ground with the help of an open cage. Down in the shaft he would instantly become aware of the darkness and the damp acrid smell of the coal dust as it entered his nostrils. Once dropped off below he would begin the walk along the gangway off the main shaft to the job location. The hike through puddles (and sometimes torrents) of brackish water could be as long as one mile.

The only sound as he walked was the crunching of rocks beneath his boots and perhaps the grunts of other miners working nearby, their tools clinking on the rocks. Given entrance through a giant wooden door that controlled the flow of life-giving fresh air by the nipper. The nipper sat all day at the door alternately opening and closing it as ordered to keep a constant supply of air inside the mine. He would report to the fire boss, the man responsible for the mine's safety. It was he who was first in the mine each day to determine bare minimum safety conditions for the miners to work in.

First he checked for adequate fresh air and to make sure the ever-present deadly black damp (methane) gas was not a threat. Along the way he prodded the roof of the shaft for signs of instability. His hammer sounded "ping" if solid and "pong" if the area overhead was ready to give way. The perils in the mines were many and they had to be constantly guarded against. Explosions could occur instantly. Flooding was always a possibility. Cave-ins would mean you might be buried alive; you were considered lucky if you were killed outright. My father, a coal miner, used to tell me how he and other miners would feed the rats inside the mine from their lunch pails. The presence of the rats, he said, assured them the conditions were safe. The sight of the rats hurrying for the surface struck fear into the miners' hearts.

The end of the breast of the mine—the main gangway—was called the "face." This is where the miners did their work. Heath would have carried all the tools he would need over his shoulder on the way to the face. Picks, shovels, drills, axes, lumber, and explosives were hauled in on the strong shoulders of the miners. In the noise and the dust he and his fellow miners would get to work bracing up the ceiling and boring holes into the rock. Explosives would be tamped into the holes and the charges set off. Without waiting for the air to clear they would return with pick and shovel to pry loose the rock and coal by the light of their headlamp. The coal would be loaded into buggies that would be hoisted topside by teams of mules. William was paid by how many buggies or cars he got topside in a shift. Day after day the young-immigrant-turned-U.S.-citizen would toil away in the anthracite mines of Schuylkill County while living in a place much like the one shown in exhibit 9.

One fascinating job inside the mine William Heath may have experienced before being allowed to go to work as a full-fledged miner at the face was that of a runner. This worker was usually a teenager. The job requirements called for lightning quickness of both hands and feet as well as superior eye-hand coordination. The responsibility of this runner was to control the speed of the fully loaded coal cars inside the mine. The cars, when full, weighed tons. Mules hitched to the cars pulled them up inclines often as steep as 10 percent. When the cars reached the top of the incline the mules would be unhitched while the cars were still moving and turned off into a crosscut (an exit ramp off the one the cars were on) so the load could freewheel down to the next incline. As the cars picked up speed the runner had to race alongside, placing wooden sprags into the wheels to slow the cars down.

The task was an exciting challenge filled with dangers, especially during the second half of the incline as the cars began really rolling. Great skill was necessary to keep up with the cars: speed, dodging low overhead beams, hurdling piles of debris, squeezing through tight fits in the tunnel, and plodding through a foot or more of water en route. Sometimes

Exhibit 9. Aerial view of Eckley, Pennsylvania, a preserved miners village that is now a tourist attraction. (Courtesy Dan Serfass)

Exhibit 10. Coal cars and miners.
(Courtesy Dan Serfass)

things went bad right from the start. If the runner was unable to get the mules unhitched they would be pushed forward and crushed to death. Other times it was the end of the run that brought disaster. If the runner was unable to keep up and get the sprags in place, the cars rolled uncontrollably and jumped the tracks, dumping their precious cargo. Understandably, the miners courted good runners and those who were mistake-prone did not last long in the black hell. (See exhibit 10.)

Prior to the turn of the century, when William Heath worked in the mines, there were no laws governing safety standards or much of anything else for that matter. The workday had no limits. There was no such thing as a work condition too dangerous for the men to toil in. Unsafe conditions were merely a fact of life and were to be endured. Noise, dust, foul air, poor lighting, and life-threatening hazards were part of the miner's existence.

Life was cheap and the coal company's only vested interest was in getting as much black gold out of the ground as it could. If men lost their lives in the process there would be others to take their place. It was not uncommon for the company to unceremoniously discard the body of a dead miner on his front porch, leaving his grieving wife without so much as an explanation. Sometimes the miners themselves were their own worst enemies. Although they were safety-conscious and were no fools with their own lives, greed sometimes cost them dearly. In an effort to get out one more load before quitting, shortcuts were taken or not enough care given, frequently resulting in a sudden fatality.

The death rate was appallingly high. When the colliery whistle blew in the middle of the day the entire town knew that tragedy had struck once again down below. Everyone rushed to the mine's opening; the women anticipating who among them were now widows. On September 6, 1869, at Avondale just beyond the county line, an explosion and fire rumbled from deep in the main shaft of a mine. The flames roared to the surface, setting fire to the breaker, which was built directly over the shaft. For three days all hands above ground fought the fire. When they finally were able to get into the mine and reached the three-hundred-foot level, all 108 men and boys were found dead. Such was William Heath's world in the 1870s.

The miners of Heath's day were an independent lot with no love lost for the company. If one of the bosses down below became too overbearing, the men would lay down their tools and walk away from the job. Yet, they were quite extroverted, spontaneous, and as fun-loving a group of individuals as could be found anywhere. They worked hard on the job and drank and gambled equally as hard when away from work. Being surrounded by this atmosphere must have been a cultural shock to new citizen Heath.

Until the mid-1870s, when he was in his mid-twenties, Heath's life was rather uneventful. Married for several years he was the father of two sons. John was born in 1873 and William arrived in 1875.* Life for Heath was moving along in an orderly fashion. However, there were storm clouds on the horizon for Girardville and all of Schuylkill County.

During the two decades preceding the 1870s, almost one million persons of Irish and English descent immigrated to the United States. Thousands of them came to settle in Schuylkill County. The horrors they suffered on board ship were matched by the treatment they often received by port of entry officials when they arrived in America. Frequently they were hassled, separated for quarantine, and then dumped on the streets to fend for themselves. Work was difficult to find in the new land, but they were always welcome in Schuylkill County where the mines awaited them.

In coal country the company owned *everything*. The land, streets, homes, schools, churches, and stores were all company property and any use of them was at the company's discretion. The proverbial company store was the only source of goods for mules around and it stocked all the miner's family needs. Miners and their families could purchase items on credit. When payday came at the colliery, deductions were already taken out for the debt owed. It was not unusual for miners to discover that they, in fact, owed the company money well beyond their pay packet due to the

---

*A summary of all of Heath's children and proof of their births are provided in chapter 8.

inflated prices charged for company store goods. The company-owned houses were little more than shacks. Most families wore ragged hand-me-down clothes and had few luxuries of the day to delight in.

Even the law was owned by the company. The coal companies hired their own policemen who operated with complete authority. They had and used the power to enter anyone's home at anytime and make arrests without warrants. They were called the Coal and Iron Police. Their real purpose was to deal with the growing number of labor disputes initiated by the workers. The company tactics were brutal and ruthless with little recourse left to their victims.

Despite this grim picture the miners knew how to squeeze every bit of enjoyment from an otherwise depressing existence. Drinking in the taverns was by far the most popular activity to be found in any coal patch community. It was practiced with gusto and regularity by most miners. Taverns could be found on every corner and in between as well. A town of three thousand people could easily expect to boast at least thirty barrooms. Drinking was such a way of life in Girardville and in other mining towns that at one time drinking was even permitted on the job at the mines where the colliery sold it itself. It wasn't until it was deluged by angry threats from the wives of miners (in addition to realizing that it was hurting production) in the 1850s that the company finally put a stop to the practice.

Second only to drinking was the miners' penchant for gambling. A typical night in town would find taverns filled with friendly drinkers. Soon stories and tall yarns were being told with oratory-like competition interspersed with the singing of ballads. Inevitably, someone would make a statement challenged by another and sides would be chosen as hard-earned money was bet on the outcome. Little if any encouragement was needed to get most miners involved. They would wager on just about anything: pigeon races, soccer matches, cockfights, and fistfights to name a few.

In a custom brought to America from across the sea, June 25 was a holiday celebrated by all the immigrants from the British Isles. Known as Midsummer Day, it was a glorious day off from work for everyone. There were singing, dancing, and of course, drinking. Old wooden cracker barrels were filled to the brim with fireworks and set off when the day changed to night. It was foretold that young girls, if they gazed hard enough, could see the face of their future husband in the light of the fireworks.

Though the miners were a fiercely independent lot, discrimination in all its ugly forms was the common foe that began to draw the miners together in the early 1870s. Most of the miners who daily risked their lives inside the holes in the ground were Irish Catholics. Everywhere they turned, it seemed, they were met with the two Ps: Protestant power and prejudice. They found it at the mines where the owners and bosses saw to it that they received the hardest, most dangerous jobs available. They encountered it in town as well. The Irish miners lived in the worst housing on the worst side of town. They were subjected to constant rip-offs at the company store. Their religion and customs were a constant target for derision and criticism. Everywhere they went they were treated with arrogance and contempt.

In response to the mass discrimination they faced, the miners began to organize into numerous mutual aid and benevolent societies. The largest of these societies was the Ancient Order of Hibernians (AOH). Ostensibly its purpose was to promote friendship, unity, and true Christian charity among its members. In reality it was the center political and labor organization among the miners. Run by a small inner circle of secret members known as the Molly Maguires, it was facing a head-on collision with management, and William Heath was himself caught in the middle. The goal of the Molly Maguires was to attain safer working conditions, better pay, and above all, freedom from discrimination.

The Whig Party, largely made up of English, Welsh, and German Protestants, controlled Schuylkill County politics and was strongly anti-Catholic. Likewise was the major newspaper of the area, the *Miners Journal of Pottsville*. The editor of the newspaper was an anti-Irish militant and took every opportunity to defend the interests of the coal barons.

The miners sought refuge in unions. Still, early attempts to organize them failed, probably due to their fierce independence. Long before 1875 a pattern of violence and disregard for the law was growing. By the 1860s a strong wave of anti-Catholic feeling struck Schuylkill County. Labor unrest grew to a dangerous level. Bodies began to turn up everywhere. Company men were found at the bottom of mine shafts, along country roads, and in their homes. In the first three months of 1867 there were five murders and six assaults in the county. The next year John Siney (see exhibit 11) managed to organize the Workman's Benevolent Society with some thirty thousand members, most from Schuylkill County.

The unions grew in direct reaction to the harsh conditions the miners found themselves subjected to. In one of their few social outlets—the taverns and beer gardens as they were called—loud protests and heated debates about the course of action they might take were constantly heard. Sporadically and over time various attempts to pick a leader and organize into groups that sought relief were tried. By 1868, with the violence at its worst ever, Siney met with the most success by managing to enlist some thirty thousand members to the cause. A bigger piece of the pie and freedom from the Protestant oppression were the goals. Even though Siney was not successful it was an important beginning in the development of the miners union that John L. Lewis would one day command.

There are few things in this world that deliver a better wake-up call than a dead body. The owners of the coal companies were clearly alarmed. It was a dangerous time in Schuylkill County and dangerous times spawned dangerous liaisons. Management was in need of a plan to break the union and a fearless man to carry it forth.

The coal companies found their salvation in Franklin Benjamin Gowen, the son of an Irish Protestant merchant from Philadelphia. The elder Gowen made a small fortune as a merchant selling groceries and liquor (the two necessities of the times). Son Frank was well educated and possessed a definite flair for the dramatic. As a young man he moved to Pottsville to try his luck in the coal business. At age twenty-three he'd gone bankrupt. Undaunted, he decided to try the law profession. In 1860 he was admitted to the bar, a job that he was well suited for with his skillful oratory. By age twenty-six he had risen to District Attorney for Schuylkill County. He became legal counsel for the Reading Railroad Company and, after winning an important case against the Pennsylvania Railroad, was catapulted to its president (see exhibit 11). The Reading Company was started in the 1820s as the Philadelphia and Reading Railroad by none other than Steven Girard. One of the nation's richest men, it was Girard who forsaw the economic future of developing the coalfields of Schuylkill County. At first it struggled because of the refusal of the state legislature to grant the railroad the right to operate coal mines. There were times when the company's bonds sold for under 75 percent of par and creditors came to Schuylkill County to seize locomotives to try to regain some of the money they invested. Finally, in 1868 the law was changed, allowing the railroad to expand into coal opera-

Exhibit 11. [*Left*] **John Siney, leader of the Workingmen's Benevolent Association of St. Clair, the first effective union of anthracite miners, pictured in 1868.** [*Right*] **Franklin B. Gowen, president of the Philadelphia and Reading Railroad and the man who crushed the Mollies.** (Both courtesy of the Historical Society of Schuylkill County, Pottsville, Pa.)

tions. The timing and location were perfect for Frank Gowen, who convinced the English banking firm of McCalmont Bros. of London to invest $40 million buying up over one hundred thousand acres of prime coal land in Schuylkill County. The move reversed the fortunes of the company making it, during its heyday, the largest corporation in the world.

In 1871 labor unrest was heating up dramatically. A strike in the county coalfields resulted. It was certainly not the first strike in the county. On July 7, 1842, the first strike was called after a meeting of miners in the Town of Minersville. Frank Gowen vowed he'd turn Schuylkill County into a howling wilderness before giving into the damned miners. The strike was crushed by the crafty moves of Frank Gowen. During 1873, in a daring move, he met with the other coal owners in the industry and together they fixed the price of coal at an average five dollars a ton. He also carved up the market, seeing fit to give the Reading Company 28 percent of the pie. In addition he saw to it that the other mine owners held the line whenever a strike would break out. Now, with all his management ducks in a row, Gowen was ready to take on labor.

The state legislature began an investigation into his price fixing, which was not technically illegal at the time. But in another bold move Frank Gowen neutralized the state's investigation of his activities by charging the Molly Maguires with being responsible for everything that was wrong in the coal industry. The secret society, as he called them, was blamed for conducting political and industrial sabotage. Due to the Mollies' random acts of destruction and violence, public sentiment was weighted against them. Fearing lack of public support the investigation was dropped.

The president of the Reading Company was on a roll. As the most powerful man in the county he continued to play his strong cards. It was he who conceived and put into practice for the first time the concept of criminal liability in a labor dispute. It was a brilliant legal move demonstrating to all that he was the most formidable opponent the miners would meet. Gowen's theory was that under Pennsylvania and federal law property (i.e., the mines) was a vested interest of the coal companies. As such, the owners of the coal companies had the God-given right to protect their property. Therefore anyone who attacked company property, its employees, or the police hired to protect its property was breaking the law and could be moved against with whatever means were deemed necessary.

Having firmly established this maxim in the courts Mr. Gowen set about strengthening his position. He beefed up the number of Coal and Iron Police. Instructions were given to them to shoot to kill for, he told them, maiming an opponent only makes him angrier and incites the rest to greater violence. With his newfound power, Gowen went for the very throats of the Molly Maguires. He paid the famous Allen Pinkerton $100,000 (an immense amount of money for the time) to infiltrate the secret inner circle of the Mollies and gather evidence of their illegal activities. Knowing full well where all this was going to lead, Frank Gowen made his final move. Calling a meeting of all the other mine owners he coerced them into joining the Schuylkill Coal Exchange. As members he had them agree to reduce the miners' wages by 20 percent. In anticipation of the announcement, the owners began stockpiling coal to weather the storm they knew was coming. The head of the Reading Company had left the miners with no alternative. The scene was set for what is known in Schuylkill County as the Long Strike of 1875.

Oral family history relates that during this period of time William Heath made an important career change. He left the mines and began working as one of Frank Gowen's Coal and Iron Police. He probably made this switch in late 1874 or very early in 1875, for the tax records in 1874 still had him on the roles as a miner although it is unlikely they would have taken the trouble to go into the books to enter such a minor change. An extensive search was conducted in the Reading Company archives. Due to a fire at the turn of the century that destroyed many of the records of this time period, an exact date was unable to be located.

Why Heath made this change in jobs is uncertain but there are some credible reasons for the move. First, there is the single issue of economic survival. The labor unrest at the time was intense. Miners were constantly conducting job actions such as slowdowns, sabotage, and strikes. The result of such activities meant that very few buggies (cars) were being loaded with coal and the miners' paychecks suffered accordingly. William Heath had a wife and two young sons to feed. The money being paid to the Coal and Iron Police was good and must have been tempting. After all, desperate times called for desperate measures on both sides.

Another equally convincing reason presents itself: When William Heath viewed the circumstances that existed in Schuylkill County at the time, he took stock of his political and religious convictions and almost certainly found his allegiance rested more on the side of management than with the miners. In fact, it is hard to imagine him expressing any loyalty to the miners at all. He was not Irish nor was he Catholic. As an English Protestant Heath almost certainly found common ground with others like himself in the Whig Party, the party that was staunchly pro-management and decidedly anti–Irish Catholic miner. In retrospect one could almost predict the move Heath would make.

The bitter strike of 1875 began in January. It lasted six long and bloody months. Throughout the ordeal Schuylkill County was more like a war zone than a small civilized piece of America, for war is what had been declared by Frank Gowen and the mine owners against the miners. As the undeclared conflict raged Gowen unleashed a ruthless and unrelenting plan to crush his foes. Much to the consternation of the miners, strike-breakers (called "scabs" by the miners) were brought in by the trainload. The miners tried to prevent these interlopers from taking their jobs, but they were driven back. The Coal and Iron Police (of which Heath was now a member) along with some hired thugs hunted down the most vocal of the strikers. Under the cover of night they beat the miners to a bloody pulp with their clubs. Union leaders were also stalked like animals and murdered in cold blood. The miners' union was befuddled. It seemed Frank Gowen anticipated their every move; almost like there was an informer in their inner circle. And in fact there was.

Knowing the miners to be a hardy, independent, and fiercely proud group of men, Gowen knew they wouldn't go quietly into the night. With their backs to the wall they fought back, employing the same tactics Gowen used. The Mollies hunted down company bosses and murdered them. They went on nighttime raids dumping coal cars, derailing the trains, and setting fire to collieries and the homes of company bosses. The Coal and Iron Police were hated as much as Frank Gowen himself. The police were tracked down in the same manner as the company bosses. At first an attempt might be made to intimidate the offender by nailing a threat to the man's door. If that failed he would be found dead a short time later (see exhibit 12).

There was total disregard for the law on both sides. So much so that the governor called a heavily armed militia into Schuylkill County to patrol the streets in an effort to put a stop to the almost daily murder and mayhem. Life in the county and specifically in Girardville was ugly and dangerous. The borough witnessed more than its share of violence and destruction. Girardville was the central headquarters for the Molly Maguires, and its leader was a tavern owner named Black Jack Kehoe.

Historical analysis of the economic and political climate of Heath's hometown in Schuylkill County around 1875 was a microcosm of what was taking place in the country as a whole. As William Heath was doing battle with labor unrest, economic depression, and discrimination, so, too, was the rest of the nation. His concerns must have been the very same concerns shared by thousands like him across the nation.

As America made ready to celebrate its first centennial, threats to the big bash loomed on the horizon. As a precursor of things to come an economic panic broke out in 1873. The post–Civil War era had, as most wars do, given way to deficit spending, corruption, waste, and overspeculation in the areas of mining, agriculture, and the railroads. Tight money in the

Exhibit 12. Molly death threat. (Courtesy of the Historical Society of Schuylkill County, Pottsville, Pa.)

financial centers of Europe caused many of the bankers on the Continent to call in loans made to investors in the United States. As more and more investors defaulted on their loans the financial situation in the country began to unravel. Shockingly, the giant New York banking firm of Jay Cooke & Co. filed for bankruptcy. Repercussions were felt everywhere as boom turned to bust. Thousands of businesses from small towns to the big cities failed. In New York City a virtual army of unemployed panic-stricken people rioted, fighting the police, in search of restitution. Other cities in the East, like Pittsburgh and Baltimore, experienced the same civil unrest. Eventually, federal troops were called out to quell the rioting.

Even though the nation was on the doorstep of the coming Industrial

Revolution money was in very short supply. A series of scandals rocked the country, as it appeared that nearly everyone had their hands in someone else's pockets. There was a failed scheme by some investors to corner the gold market. Ties to this plot led to the White House as President Grant's own brother-in-law was implicated. The infamous Boss Tweed in New York City was exposed by the *New York Times*, accused of siphoning off hundreds of millions from the operation of that mighty metropolis. The Mobilier scandal revealed the greed that found its way into the halls of Congress. A crafty spin-off from its parent company, the Union Pacific Railroad, the Mobilier contracted to build the line out west. They paid themselves thousands more per mile than it cost to lay the track. In addition they stood to make exorbitant amounts in stock options. One year the corporation paid an unheard-of dividend of 348 percent! An investigation revealed a number of key congressmen holding shares of this lucrative stock.

Soon the country learned of two more black eyes on the faces of Washington officials. A huge whiskey ring was uncovered. Through devious bookkeeping it had been bilking millions of dollars in revenue from the U.S. Treasury on the sale of alcohol. Again the scandal besmirched President Grant. His very own personal secretary was deeply involved but escaped conviction due to the personal intervention of the president himself. One final example is Secretary of War Belknap who was caught red-handed taking a kickback of $24,000 in exchange for granting authority for certain individuals to supply the Indians out west with supplies as called for in our treaties. No attention was even paid to the fact that the goods were shoddy (e.g., spoiled beef and moth-eaten blankets). At least one positive thing did come from all this graft and corruption. The public outcry gave rise to massive (and much needed) civil service reform. Freedom from the recurrence of government scandals was unlikely, but the reform went a long way to slow down the ugly beast.

The linking of the West with the East by rail became a priority after the close of the Civil War. Up to that time a majority of the miles of track were east of the Mississippi. Congress authorized the Union Pacific to take on the task of building the first transcontinental line. It began moving west from Omaha, Nebraska, and was to terminate in Sacramento, California. The Union Pacific was given a generous $16,000 a mile to do the job across the flat prairies and $48,000 for going over the mountains. Along the way the company was granted twenty square miles of right-of-way land on both sides of the track for each mile completed. This added up to hundreds of millions of acres of prime real estate. After completing the line the railroad sold off much of this land for a handsome profit. Thousands of Irish immigrants from England and Chinese coolies from the West labored hard at this engineering marvel, sometimes laying ten miles of track in a single day. By the time of Heath's death in the early 1890s there were five completed transcontinental lines crisscrossing the nation.

Bound together by these steel rails our nation began to grow like an adolescent boy. The railroads offered a safe and efficient way for settlers and immigrants to get west and stake out their land and mining claims. Manufactured goods now had a way to get to far-reaching markets. The miners could get the ore out and necessary supplies in. The fruits of the farmer's hard labor as well as the cattleman's beef had an unimpeded route to the buyers in the East. All along the way, from start to finish, the railroad had become the maker of millionaires.

As is usually the case, the common working man did not share as proportionally in the wealth made by the newfound millionaires. The woes of the men like the miners who toiled with Heath in the coal holes of Schuylkill County were the same woes of many across the nation. It included thousands of factory workers and railroad laborers who shared the same frustration in dealing with a large corporation. The company held all the power; working men could do nothing but bluster and bluff. Corporations had the wealth, the accompanying high-priced lawyers, federal judges, and the newspaper all on their side. They could discriminate against lockout, and fire employees at will. To thousands of workers the situation was little better than the slavery of the Blacks they had fought to free. The incentive for the growth of labor unions like the Ancient Order of Hibernians or the Workmans Benevolent Society in Schuylkill County is not hard to imagine. Still, it would be years before they became strong enough to bring about reform.

The exact date is not known to the family but, according to oral history, William Heath awoke one day to find a death threat (much like the one in exhibit 12) nailed to his door. Girardville was definitely not the place for a Coal and Iron Policeman to live. It could, however, very likely be a place to die given the dangerous circumstances that existed there.

Apparently William seriously considered the consequences of the threat and made what must have been for him a very difficult choice. He decided to leave Schuylkill County in a hurry. Heath surely must have felt that his life was in real danger to leave his home, his wife, and his children behind. Perhaps he planned to send for them as soon as possible. Somehow his plan never got that far. Most likely his family, during his absence, lived with his in-laws, the Swansboroughs. William Heath may have told them his leaving would be only for a short time; he'd be back as soon as the unrest and violence subsided. Little did he know that he was going from one life-threatening situation to a much worse one.

Considering the tension between the owner and the miners, it is easy to fathom how William would become a member of the Coal and Iron Police in 1875. Fitting into the Protestant Whig mold perfectly, his allegiance to both would have led him right to the camp that family history gives us. One can relate to why he left town so quickly. Death threats can urge one to react unlike anything else. What is much harder to understand is why Heath did not take his family with him.

Without being able to get inside Heath's head, this is impossible to answer. Perhaps he had absolutely no idea of where he was headed and felt it made no sense to drag them into the unknown. Perhaps he figured that with his wife's family around as well as his own (his brother, Arthur, and sister, Ruth) his wife and son would be much better off in Girardville than where he was going. Certainly he was going to have to travel fast and loose, and subjecting them to such harsh conditions must have figured into his thinking.

A third possibility that comes to mind (before judging him too harshly) is this: Margaret Heath was most likely with child when all this came crashing down on them. Since we don't know the precise date William left town, this is hard to nail down. However, we do know (by comparing dates given later) that Heath's second-born came into the world roughly one month after he joined the army on his way to Montana. Assuming William was the father, Margaret had to be very pregnant while Billy was packing his bags. They may have agreed he leave immediately while she, at least for the time being, remain at home. Or, William may have asked her to go with him and she refused, preferring to be with family when her time came to give birth. She may also have been fearful of slowing him down if she went along. We will never know what emotional conversation took place between this husband and wife. We can only be intrigued by the touching human drama of two people in love suddenly cast into an unthinkable situation.

Specifically when William Heath left Girardville or by what means he traveled is not known. What is acknowledged is that he began to move westward. During Heath's absence Frank Gowen emerged victorious. The miners eventually lost solidarity and begged for compromise. Gowen refused all overtures. In June the miners slunk back to work beaten into submission by the president of the Reading Company. Through the ordeal Gowen spent $4 million to crush the miners. Not satisfied he set out next to totally destroy the Mollies. In 1876 the first trial against them opened in the town of Mauch Chunk in northern Schuylkill County. In May of that same year six more Mollies were put on trial in Pottsville. All were found guilty and hanged on Black Thursday, June 21, 1877. Soon after, Black Jack Kehoe himself was arrested and eventually wore the noose around his neck.

Gowen was successful with every case prosecuted against the Molly Maguires. This spotless record was mostly due to the damaging testimony from the spy who penetrated the inner circle—the spy who Gowen paid Pinkerton $100,000 to plant. His name was James McParlan. Frank Gowen, dressed in evening clothes, tried the cases in person as the chief prosecutor for the state. He dominated the proceedings, reveling in every minute of it. His victory over the miners was complete. It would take many years before they reorganized and made inroads toward management under the powerful United Mine Workers Association.

Frank Gowen, always conducting himself as if an actor on stage, left this world with dramatic flair. In 1889, at the age of fifty-three, he was found dead by his own hand behind the locked doors of a hotel room. Questions arose and were never answered. His law practice was booming at the time. He left no suicide note to explain his actions. Suspicions ran high that somehow, some way, an old Molly enemy had found a way to get to him and did him in.

Chapter Two

A FEARFUL FLIGHT WEST

William Heath's flight west took him to Cincinnati, Ohio, where army records verify he enlisted in the United States Army on October 9, 1875 (see exhibit 13). It is unclear whether he planned to join up from the outset or if the idea came to him when he got to Ohio. If it were the latter he would not be the first man to seek adventure and fortune out west by joining the army. Many men used the army as an opportunity to get where the action was with a convenient means of transportation and some protection to boot. When the time was right they would merely go over the hill never to return. I highly doubt that adventure was the main motivation of Billy Heath. More than likely the army was a safe place to get lost in as well as a way to send home to his wife and children most of the thirteen dollars a month he would get.

The army record of enlistment shows William Heath signing his name whereas on his citizenship application three years previous he merely made his mark. Apparently William learned to read and write. Heath's physical description is given for the first time. The twenty-seven-year-old man is five feet seven and one-quarter inches tall and has blue eyes, brown

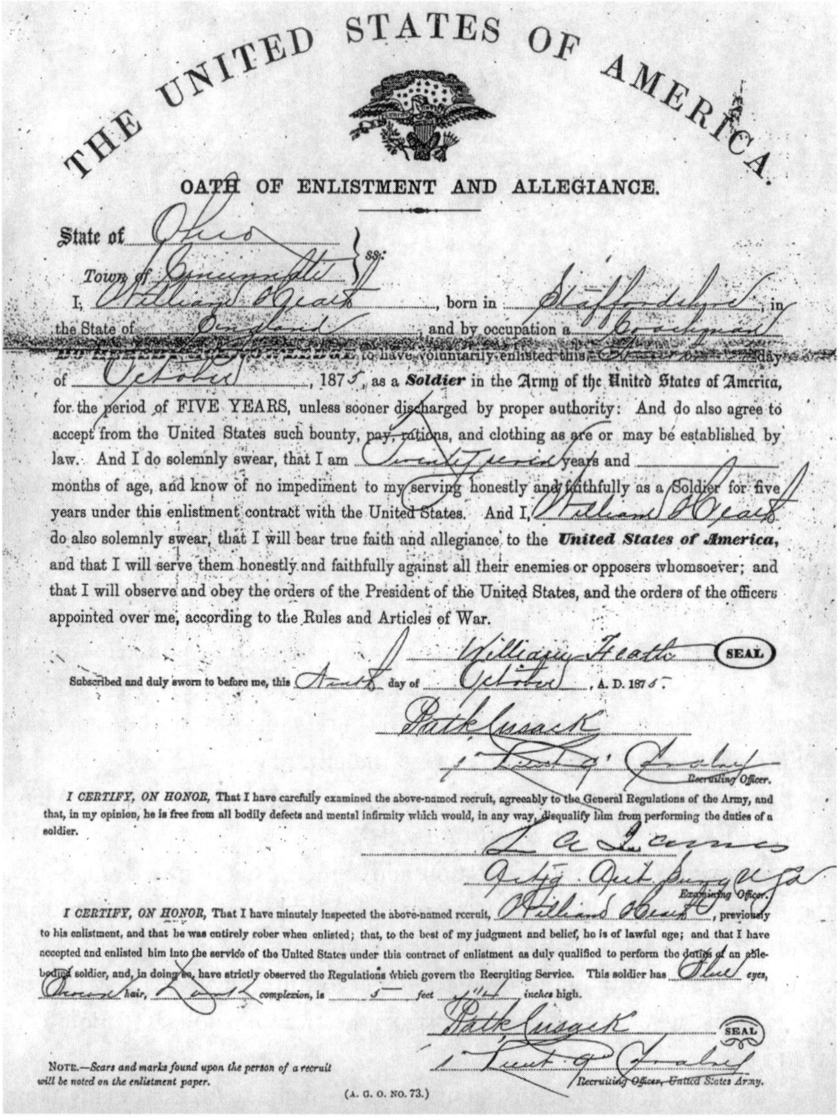

Exhibit 13. William Heath's enlistment record. (Courtesy U.S. Army Archives)

hair, and a dark complexion. No scars or marks were noted on the new soldier by recruiting officer Lieutenant Cusack. There are explanations for Heath's signature on the enlistment in comparison to his mark on the citizenship application. Either William learned how to write by 1875 or someone signed his name for him in Cincinnati. This was not an unheard-of occurrence in recruiting offices at this time. Old records are often incomplete, as illustrated by the official forgetting to sign Heath's citizenship paper at the very end despite signing it in several other places.

The significance of exhibit 13 can hardly be overstated. As per family history we now have proof that William Heath of the right age originating from Staffordshire (ergo the same Heath as the one living in Girardville) had joined the army, later to be assigned to the Seventh Cavalry. It is an important part in establishing Heath was at the battle.

The army promptly assigned him to Company L of the Seventh Cavalry, sending him on to Fort Lincoln in the Dakota Territories where the Seventh Cavalry was in the process of being beefed up in anticipation of being assigned the task of putting down the Indian uprising in the Black Hills area, now South Dakota. Young William Heath had no idea of the future to which he had just committed himself. The murder and violence he left behind in Schuylkill County would pale compared to the carnage he would take part in at Little Bighorn.

At Fort Lincoln new recruit Heath was trained in the military arts of riding, fighting from horseback as well as on foot, and taking care of his weapons. The United States Army, noting his former occupation as coachman (even though his last job was that of miner), assigned him to the task of company farrier. The farriers were basically in charge of the Seventh's mode of transportation, the horses. The company farrier would see to it that the mounts were properly saddled. If any shoeing was needed, he took care of it. Feeding and watering the horses daily was also his responsibility as was keeping a supply of necessary equestrian equipment on hand such as bits, reins, harnesses, and the like.

Horses often came up lame or injured from stepping in prairie dog holes or from traversing steep inclines. The farrier would determine the nature of the injury and treat it with the liniments and accepted practices of the time. He was also called upon to treat wounds to the animals, be they from gunshot, arrow, or snake bite. With approximately one hundred horses to care for in a company, the farrier was kept busy. It is this assignment to the job of farrier that leads to some interesting theories about William Heath's activities during the Battle of Little Bighorn (more on this in chapter 8).

As an impromptu veterinarian company farrier Heath was accountable for the general health of the horses. Without a dependable supply of robust animals the mobility of the mounted Seventh was negated. One of the most common ailments a horse could suffer was saddle sores. Inane as it may sound, saddle sores could become serious enough to prevent the horse from being ridden. This could leave a rider without a mount, putting the entire mission at risk. Wool blankets were usually used under the saddle. Sweating or excessive saddle movement could irritate the horse's back as the wool fiber rubbed the skin raw. The treatment farrier Heath might have used would have included making sure the horse's back was thoroughly dry and then applying liniment. Finally, the saddle area would be covered by a piece of burlap which was much more soothing to the animal's skin. The burlap also kept the sun and insects from exacerbating the problem. Stopping to walk the horse with saddle loosened also helped to keep the condition from getting worse.

The farrier had to be very diligent about the care of the horse's feet. Hooves had to be cleaned several times daily to keep the animal free from debris and moisture-laden material. Failure to do this could allow a fungus to take hold and totally incapacitate the rider's steed. Every several weeks the old shoes were removed and the animal reshod to keep it fit for duty. The farrier had to also combat other maladies such as shin splints, stress fractures, ear infections, diseases of the mouth, and heart ailments such as could be brought on by riding the horse to exhaustion. Telltale signs of trouble included changes in the horse's droppings, bad breathe, lack of appetite, and rapid heartbeat. The end of the day's march brought rest for most of the troopers but not the busy farrier.

The duties described, though numerous, were not the only obligations of farrier Heath. During the demanding marches and in the heat of battle he had the same responsibilities as the other soldiers. He trained with the rest of the men, practiced offensive and defensive tactics, learned how to shoot from horseback and on foot, cared for his firearms, and went hand to hand with the enemy if necessary. He may have been a farrier but he was still first and foremost a mounted member of the proud Seventh Cavalry.

When under attack Heath could just as likely be called upon to guard the perimeter from the enemy's attack, be part of a scouting team sent out to gauge the foe's movements, serve as a messenger, or go along as an escort with a runner. He endured the same long hours in the saddle; suffered the same exposure to blazing sun, bone-chilling cold, and icy-cold rain; ate the same meager field rations; and choked on the dust just like everyone else. In a tense battle situation he would be found on one knee firing at the Indians alongside his companions. Except when pulling his double duty as farrier, he was simply another fighting member of the Seventh's Company L.

Life on an isolated fort in the western wilderness could be extremely boring at times. Up with the sun to hoist the flag, the soldier's day could be filled with dull, routine tasks. The drilling was repetitious. Target practice was a welcome relief. Gen. George Armstrong Custer himself usually had two practices a day. The privilege of being excused from guard duty was afforded to the men who attained the skill of marksmen. Guard duty was a necessary evil at a frontier fort. The Indians presented a clear danger both day and night, and guards had to maintain vigil lest the hostile red men try to set fire to the post or steal the stock.

There were many civilian jobs that fell on the soldier's shoulders as well. Enlisted men like Heath could be used to cut wood, serve as teamsters driving supply wagons, hoe gardens, and work as quarrymen. They might spend day after day building roads or constructing some public building. No doubt it was welcome duty to get assignments like scouting escort duty—assuring safe passage of trains and wagons of settlers—or patrolling roads and trains. Here, occasionally, things would be livened up by a chance skirmish with some Indians or maybe even some buffalo hunting. At night they could lie awake on their bedrolls looking up at the stars, enjoying nature's show from the stark blackness below. As the embers of the cook's fire died out, the only sounds heard were the comforting footsteps of the sentry and the rustle of the horses. In times of peace it wasn't a bad way to live.

When Heath arrived at the Seventh Cavalry headquarters in the wilderness fort at Lincoln, he found himself amid a hodgepodge of personalities and military résumés. The Seventh was a diversified collection of social and cultural lineage. The first headquarters for the Seventh was Fort Riley in Kansas when the regiment was newly formed in 1866. Under the jurisdiction of the Department of the Missouri Territory, it was assigned to frontier duty. Its mission included protecting farming communities in the area, guarding the Santa Fe Trail, and watching over the railroad construction crews pushing the tracks west. The unit's high commander was Gen. William Tecumseh Sherman, remembered for his famous Civil War march to the sea and second only to Grant in the respect accorded by a grateful nation.

The Seventh Cavalry that Heath joined was an odd mixture of seasoned veterans from both the Civil War and Indian Wars plus green recruits like William himself. The experienced members had ten or more years under their belts. Some of them were West Point graduates. Many of them had already proven their mettle under battle conditions. The new enlisted men included a large number of immigrants from England. Historians have noted that often they were misfits avoiding the law or adventure seekers looking for some pot of gold. The educational level among their ranks was low. Many, unlike William Heath, could not read or write.

The Seventh's mission lacked the clear-cut patriotic backing that grounded the actions of military units of the Union Army during the Civil War. This absence of purpose, along with the often boring nature of army life when not in battle, contributed to heavy drinking and gambling among both officers and enlisted men. The average age of most troopers was twenty-seven years—William's age.

In spite of popular belief, General Custer was not held in high esteem by all in the Seventh Cavalry. Unfortunately, factions permeated the unit. Capt. Frederick Benteen led the anti-Custer feeling and the dislike between the two men was well known. Benteen was from Virginia but chose to fight with the Union in the Civil War. He was five years Custer's senior and had been with the Seventh Cavalry longer than Custer was. Maj. Marcus Reno was second in command to the young General Custer. He had little Indian experience even though he'd been with the Seventh since 1869. Reno once, in Custer's absence, tried to take over command of the Seventh. Ever since that time the relationship between Reno and Custer had been strained. Everyone at Fort Lincoln knew there was no love lost between them. Seldom was there found indifference to the dashing boy general. You either hated him or worshiped him. The officers and enlisted men of the Seventh fell into one of these two categories according to their whims. Custer recognized this divisiveness immediately after taking command as can be seen in his letters to his wife, Libbie. He thought that after they had experienced the rigors of battle together they would acquire the unit cohesiveness that he wanted. Sadly, that never happened.

Maj. Marcus A. Reno [*left*] **and Capt. Frederick W. Benteen** [*right*].
(Both courtesy *Dictionary of American Portraits*, Dover Publications, Inc., 1967)

The enemy that Heath and the Seventh Cavalry were to meet, the Plains Indians, had a long and storied history. They were a proud people; a people never quite understood by the white man. The history of an important event like the Battle of Little Bighorn is most often penned by the eventual victor. The winner's point of view dominates the literature while the point of view of the loser tends to get brushed aside. History books often convey more about what they don't tell than what they do. Buyers of books like to read the victor's side and not the vanquished's. Most past accounts of Little Bighorn have catered to the patriotic fervor of our nation's white majority. Some attempts have been made to depict the role of Native Americans with less prejudice. However, the truth is that the dearth of literature published on the battle is from the white man's point of view. In telling the story of William Heath I have made every attempt to research as many points of view as I could find, whether they be Native American or white.

The Sioux Indians of America were a powerful people whose domination was as far-reaching as any other tribe. These Indians ruled the territory west of the Mississippi River from the Canadian border to as far south as Louisiana long before the time the first French trapper appeared in the area. The best known of the Sioux tribes to the white man are the Dakotas. They are the ones so often conjured up in the imagination when someone speaks of Indians. Images of the slightly built body with delicate facial features punctuated by the prominent nose come to mind. Historians, to illustrate their descriptions, point toward the Indian-head nickel as the perfect example.

Great chiefs come to mind when considering the Sioux. Chiefs like Crazy Horse, Sitting Bull, and American Horse sitting around a campfire with their eagle-feathered headdresses or smoking red clay pipes are part of our vision. In the procession leading to the inauguration of the president of the United States, there are always a few Indians on horseback, wearing the traditional Dakota costume.

The word "Dakota" means friends. Not unlike many other far-flung tribes, they had numerous dialects in their language. Despite this they could easily converse with one another no matter how far removed. It is said that their native language is the most pleasing to the white man's ear out of all the tribes. In general, the Dakotas were friendly to each other and seldom made war among themselves.

In terms of sheer numbers, the largest segment of the Dakota tribe is made up of the Tetons. They were subdivided into the Seven Tribal Council Fires made up of the Blackfoot, Hunkpapa, Oglala, Brule, Miniconjou, Two Kettle, and Sans Arc. Over the years of conflict with the white man the Oglala were the most successful. They had the largest numbers and were known for training fierce warriors. For hundreds of years they lived a nomadic life, finding sustenance from the flesh of the buffalo. These great beasts supplied the Indians with everything they needed to survive. From the animals they made tepees from the skins, ornaments for their bodies, tools to work with, arrowheads and knives to fight with, as well as spoons, rattles, ropes, thread, bags, moccasins, and clothes. Their whole existence was centered around this big shaggy beast.

Understandably, the buffalo took on magical qualities to the Indian and was a major part of his religious beliefs. Before a hunt the tribal leaders prayed to the great-unseen buffalo for a plentiful harvest, always taking great care to slay only what they needed. They feared any wastefulness would anger the Great Buffalo God, bringing starvation upon the vil-

Buffalo hunt chase. (Courtesy Denver Public Library, Western History Collection/George Catlin/TS#N6)

lage. After a kill their first order of business was to give thanks to the fallen animal for giving up its life that they might continue to live. Thus, the Native Americans were this country's first consummate conservationists.

The thundering herds of buffalo proliferated in America for centuries. The first white men, the Spanish, called them hunchback cows. As late as the end of the Civil War there were still around fifteen million shaggy buffalo grazing on the open range. Once, in 1868, a Kansas Pacific train had to wait for eight hours for the herd to finish crossing the tracks before it could continue! Previous attempts to merely run through them resulted in derailment. Buffalo Bill Cody is reported to have killed over four thousand animals in an eighteen-month period while working as a scout for the Kansas Pacific Railroad.

As the railroads continued to crisscross the plains the slaughter of the buffalo accelerated like a runaway locomotive. Buffalo robes were the fashion rage and many of the animals were slain to supply this demand. The rest of the carcass was left to rot in the sun as food for the vultures and other scavengers. Limitless thousands were picked off from moving trains by enthusiastic "sportsmen" who fired out the windows for whatever satisfaction it provided them. Sadly, by 1885 it is believed there were only about one thousand buffalo left as they came dangerously close to total extinction.

The unabated slaughter of the buffalo by the white man not only robbed the Sioux of their primary food supply; it also was a direct attack on their very religion. In a two-year period in the early 1870s almost four million head of buffalo were killed. Of this number, the Indians took 150,000 to maintain their survival while the rest were cut down by whites mostly for sport. According to Dee Brown, in *Bury My Heart at Wounded Knee*, when a small group of concerned white citizens approached General Sherman in Texas about the carnage, he answered in fine oxymoronic fashion, "Let them kill, skin, and sell until the buffalo is exterminated, as it is the only way to bring lasting peace and allow civilization to advance."

The Sioux had an unbridled love affair with horses. Horses were a sign of wealth and power among tribal members. There were thousands of wild ones roaming freely on the plains; however, the Indian preferred to steal his mounts from neighboring tribes. Not only were the horses already broken in, it was a rite of passage for a young brave to demonstrate his courage by robbing the prized ponies from an opponent. The Indians started this practice first with the Spanish, and then, in succession, with the French, English, and finally the Americans. Meticulous records were kept on individual exploits much like a general's medals or a scholar's degrees. If discovered, the thief would fight to the death for the horses, bringing back a scalp to back up his tale of bravery. Oddly enough, the white man introduced scalping to the Indians. The French trappers used scalps as proof of payment to friendly Indians for having eliminated an unfriendly one.

The Indians learned to be one with their horses. The white man had never before seen the feats of horsemanship demonstrated by the Native Americans. While riding at full gallop in battle braves could shoot with great accuracy by standing high in the saddle (most Indians did use saddles). In close quarters they would use the horse as a shield and swing around and under the horse's belly to keep up fire. A favorite tactic was to flee from the white man into the tall grass of the plains and appear to fall from the horse. If the white man foolishly followed, Indians sprang up from all sides and quickly dispatched him.

The Tetons were well known for practicing a sensational religious ceremony of self-mutilation. During the annual great tribal festival known as

Sioux Sun Dance. (Courtesy Denver Public Library, Western History Collection/George Catlin/X-33637)

the Sun Dance, only a handful of braves would attempt this dramatic feat. A sharpened stick was forced through both sides of the flesh of the chest and arms. A well-made cord was tied to these sticks and fastened to a stake driven deep into the ground. The participant danced, chanted, and prayed to the gods to provide him with the power to break free. Then, he would run full speed to the end of his tether, ripping himself free by literally tearing chunks of flesh from his body. It was considered a good sign if he passed out for, while unconscious, it was believed he could talk with the gods and foresee things to come.

This ritual was to become prohibited on the reservation. The white man considered it barbaric. He also noted that when practiced by the Indians the entire tribe became greatly agitated and extremely difficult to control. Clark Wissler, in *Indians of the U.S.*, noted that it was during the Sun Dance festival in 1876 that Chief Sitting Bull reported having a vision from the gods. In his revelation Sitting Bull foresaw the Battle of Little Bighorn and clearly witnessed the total defeat of the white soldiers a few days before the battle. His vision strengthened the bravery of the warriors who followed him.

The Dakota Indians, as well as others, often traded for arms. It was not unusual to find them better equipped than their U.S. Army foes. They frequently possessed the newer repeating rifles and were expert marksmen with them. Under government treaties the Indians were given ammunition ostensibly for hunting. Instead, they stockpiled it for war, preferring to hunt their game with bow and arrow. Over the years the Sioux tribe grew in power and waged war on neighboring tribes, mostly the Crows. There was little trouble with the white man prior to 1849. The California Gold Rush of that year brought about a change in the situation. Thousands of white men in wagons filled the trails west starting the conflict between the peoples. The Treaty of Fort Laramie in 1851 brought temporary peace. At this time a new Teton leader came onto the scene. Red Cloud was a self-professed warrior not given to oratory and negotiation. There was bloodshed in Nebraska and Minnesota. Red Cloud was suspected as the ringleader of the uprisings. As a result federal troops were sent to the Wyoming and Montana Territory, setting the stage for Little Bighorn but I am getting leaps and bounds ahead of myself.

As crucial as it was to understanding the conditions in Schuylkill County prior to and around 1875 in looking at William Heath's life, so, too, is it necessary to grasp the circumstances of the Indian situation leading up to the Battle of Little Bighorn.

In a country as young as the United States a true sense of identity is not nearly as clearly defined in as many other countries. The scant number of years we have existed as a nation plus the melting pot makeup of our population makes this so. Nations like China, Greece, and England took thousands of years to develop their customs and traditions. Our so-called traditions, despite the wishes of our historians and scholars, fade when compared to those of so many other cultures. If, in fact, we do have an ancient identity it is in the Native Americans. By virtue of their many centuries of inhabiting this country, they are our common identity.

Since the first white man set foot in America legends surrounding the Native Americans sprang forth. Some embellished and many tarnished these people depending on the whims and ulterior motives of the white man. In order to justify these motives we've signed treaties. Lots of treaties. Over four hundred of them in truth. One would be hard-pressed to locate one that we have kept in good faith. Sometimes it took many years. Even so, the result was always the same. A case in point: In 1794 George Washington himself sent Timothy Pickering to upstate New York to negotiate a treaty with the Iroquois Indians. The government swore to respect the boundaries of the Indian lands, never to disturb the Indians' use of the land, and never to lay claim to the same. It took nearly two hundred years until, in the early 1960s, the government built a huge dam on the Seneca reservation in upstate New York, flooding most of the reservation.

The justification for such action originally began with the "Doctrine of Discovery." Strangely enough this brilliant idea was the invention of the European Christian Church. Simply stated, it negated all rights of the natives of a country, taking away their title to the land, and forced them to sell that land to whichever country "discovered" it first. In the eighteenth century when the Native Americans balked at this unusual practice of the white man we extended the doctrine from purchase to one of purchase or conquest. All this in the name of God.

The history of the Indian Wars revolves around this doctrine and its broken promises. The Black Hills of southwest South Dakota was the center of the universe to the Sioux Indians. The Crows were the first of record to reside there, enjoying this land of plenty. By 1770 the Sioux had wrestled the land from the Crow and driven them out. The Indians had good memories and subscribed to the philosophy that any enemy of an enemy of theirs was their friend. This is why so many scouts in the United States Army during the frontier days were from the Crow tribe; they had a grudge to settle with the Sioux who wandered down to the Black Hills from the headwaters of the Mississippi.

Having been to the Black Hills (while enroute to Little Big Horn) I was surprised to find that they are not black but more a combination of grayish blue mixed with dashes of red. The Indians were drawn to this region for two reasons. First, it literally teemed with game of all kinds. Large numbers of deer, elk, antelope, beaver, and fox could be found. This abundance was seldom found anywhere else by the nomadic Sioux. Second, because of the bountiful game and the strange rock formations in the area, the Black Hills were considered sacred ground to the Indians. They came here to pray to the gods. The unusual rock shapes were the steeples for their church. The white man's desecration of the hills was no less shocking to the Native Americans than a swastika on the wall of a synagogue.

The Sioux (or Teton Lakotas) formed a loose and mostly peaceful alliance with their neighbors the Cheyenne. They shared the abundance in the hills as well as treating the land as sacred and hallowed ground. Even in spite of this horn of plenty the Sioux did not stay in the Black Hills all the time. They were a tribe with large numbers of people. If they stayed in any one area for too long they could strip the land bare. They recognized this instinctively and shrewdly continued a nomadic way of life throughout their history. During the warmer months they would follow the game wherever it moved. In wintertime they sought shelter in the lower milder elevations in valleys and along riverbanks to seek protection from the harsh cold.

When the first white men passed through the Sioux territory there was no trouble between the two races. The red men viewed the whites as travelers merely passing through on their way to somewhere else. As long as they were not staying the Indians proved helpful and downright hospitable to whites. The American Fur Company built a trading post near the Laramie River in Wyoming in 1841. Here the business of trading with the Indians for valuable furs was conducted. Whiskey was introduced at this same time. The arrival of alcohol turned many Indians into alcoholics to say nothing of its influence on their decisions to take to the warpath.

The Sioux were known as brave and fearless warriors. They had a very well developed warrior organization within the tribe. There were several levels of progression starting with young brave. Accurate records were kept of each rite of passage. The military-like society coupled with their large numbers made these Indians a powerful opponent, something the white man never truly grasped. Many of the adult warriors stood six feet tall, an unusual height for Native Americans of the day.

The blazing of the Oregon Trail opened up new lands for the white settlers and adventure seekers. It began in Independence, Missouri, and ran through Wyoming and over the Continental Divide. The Indians failed to realize how many whites there really were to their east. By the time they did sense the risk it was too late. As the whites came, they drove off and decimated the game animals. The Indians reacted by attacking. In 1845 the Indians, after being showered with gifts from wagons loaded with goods, agreed to allow the trail to be kept open. The trading post became Fort Laramie, and an uneasy peace pervaded.

With the pressure brought on by the influx of so many whites over the next couple of years, the relationship between the two races deteriorated.

In September of 1851 an estimated ten thousand Indians gathered at Horse Creek some thirty-six miles from Fort Laramie. Faced with a common foe, the Sioux, Cheyenne, Arapaho, Gros Ventre, and Arikara tribes came together. Here their chiefs met with Col. D. D. Mitchell, the official representative of President Franklin Pierce in Washington. Following some old-fashioned haggling the United States agreed to leave the Great Plains for the sole use of the Indians. It also promised to give them $50,000 worth of supplies each year for a period of fifty years. The ink was hardly dry on the Fort Laramie Treaty of 1851 when the government arbitrarily reduced the number of years. Also, in reality, the actual supplies delivered were nowhere near adequate to take care of the Indians' needs.

On August 19, 1854, the Indian Wars received its official jump start near Fort Laramie. A wagon party of passing Mormon settlers was in the area taking a break from the trek west. An unnoticed cow strayed away from their campsite and wandered into a nearby Indian settlement. The Indians, surprised by their newfound good fortune, promptly butchered and ate the animal. The Mormon owner soon discovered the fate of his cow. He demanded the chief compensate him for his loss to the tune of twenty-five dollars. The chief counteroffered ten dollars and things were at a stalemate. The Mormon appealed to the U.S. Army at nearby Fort Laramie.

The next day the army sent a brash young lieutenant to deal with the matter. John Gratten was an inexperienced officer eager to make his mark on the frontier. He hurried out to the Indian camp to confront the chief. Gratten demanded that he turn over the offender for punishment. The chief refused to comply. The discussion escalated into an argument between the two and an aggravated Gratten, despite orders to avoid a fight, shot the chief dead. More shots were exchanged, resulting in Gratten and all twenty-nine of his contingent being killed.

One of the young braves on hand to witness the entire event was a twelve-year-old boy with the name of Curly. In a few years he would come into his own as the great chief Crazy Horse. Compared to most Indians, the adult Chief Crazy Horse bore little resemblance to a typical man of the red race. He stood five feet eight inches tall. Unlike most Indians he had a light complexion, light sandy-colored hair, and lacked the high cheekbones that the Indians usually had. The name Curly was given him for the wavy locks he had instead of the usual poker hair. He looked more white than red. Crazy Horse was not much of a talker; he preferred action to talk. That tendency was what made him into one of the most feared warriors of his tribe.

The Sioux Indians, after the Gratten massacre, began terrorizing white settlers and travelers in earnest. They attacked all along the Bozeman Trail as well as the Oregon Trail. The Bozeman Trail began in Missouri, deviating from the Oregon Trail at Fort Fetterman, and ended in Virginia City, Montana. Killings and raids were conducted by the Navajo in Arizona, the Apache in New Mexico, the Sioux in Minnesota, and the brothers of the Sioux, the Cheyenne, in Colorado. The war was on full speed by now. In 1855 Gen. William Harney, in retribution, destroyed a band of 136 Indians, which led to a reaffirmation of the Treaty of 1851. The Native Americans withdrew from the area and peace settled in once again. But by 1862 gold was discovered in western Montana. The white man surged along the Bozeman running from Fort Laramie to Virginia City, Montana, causing the Indians to once again take to the warpath. Congress aggravated the situation in 1865 by passing legislation allowing for the construction of a wagon road west, along with the establishment of a military fort along the Powder River in southeast Montana.

The wiley Chief Red Cloud had been stalking Col. Henry B. Carrington at Fort Kearny for weeks. Fort Kearny was in southeastern Wyoming along the Oregon Trail. His plan was to harass the officer into pursuing him while a large force lay in ambush. Finally, in June of 1865 the plan worked. One day an Indian attack on a train necessitated relief from the army. Carrington gave into the pleas of a young inexperienced captain to serve the duty. Capt. William Fetterman was disgusted with Carrington's lack of offensive force against the red men. He openly criticized his superior officer's conduct. The brash officer boasted he could whip the entire Sioux nation with just eighty men.

When Fetterman arrived on the scene, the Indians fled according to Red Cloud's plan. Captain Fetterman, despite orders not to pursue and only escort the train back, gave chase for about two miles. Red Cloud sprung his trap, killing Fetterman and all eighty men with him. On more than one occasion the use of such tactics rewarded the Indians with victory over a seemingly superior force. Indian ponies were hardy and could survive on prairie grass. The Indians themselves were quite capable of for-

The Battle of the Washita. (Courtesy Denver Public Library,
Western History Collection/Kappes/X-33801)

aging on the run. The army, on the other hand, had to transport hay for its
horses plus food and ammunition for its men. Any large-scale operation
would be severely limited in its mobility by the necessary supply train.
When Indians fled through especially rugged terrain, the army usually
found it impassable for the wagons and had to break off pursuit.

In the winter of 1868 Gen. George Armstrong Custer, stationed at
Fort Riley, Kansas, and just back from a court-martial and one-year suspension (details on this in chapter 4), joined General Sheridan for a campaign against the Cheyenne along the Washita River in western Oklahoma.
Tracking the Indians to their winter camp he attacked a sleeping village at
dawn. Chief Black Kettle was slain along with about one hundred Indians,
many of them women and children. The atrocities by these white men
were reported by Indians.

That same day Custer committed a second major faux pas. Maj. Joel
Elliott gathered a band of troopers and took off in pursuit of the fleeing
Indians. When Custer had mopped up and was preparing to leave he was
informed that Elliott was missing. Custer searched halfheartedly and left.
Later it was learned that Elliott and all nineteen of his men were killed
only a few miles away, easily within the reach of Custer.

The battle was an ugly black mark against the general. The Indians
claimed there were more women and children killed than warriors. The

soldiers, under Custer's orders, burned the tepees; destroyed all weapons, clothes, and saddles; and then cut the throats of around nine hundred Indian ponies. Some deep rifts were opened up in the Seventh Cavalry that day and they would never heal.

Back in 1864 the news of the Sand Creek massacre of Indians by whites in Colorado Territory resulted in outrage by Christian missionaries. Soon, Congress called for an official investigation. Although nothing was done to the participants, the official report of the government actually condemned the poor handling of the entire incident. The Joint Special Committee called it a scene of murder and brutality. The committee acknowledged that women and children were indiscriminately slaughtered and that these particular Indians had already declared their willingness for peace. (Similarly, the day before the Battle at Washita Black Kettle had been to Fort Cobb in southwest Oklahoma where he accepted peace. When Custer's attack came it had been in full view of a white flag hanging from a pole in the Indian village.) The final statement of the government criticized what it called acts of barbarity of the most revolting nature that were allowed to continue by those in charge. The Battle at Washita resulted in no similar outrage and demonstrates a hardened view by the white man regarding the Indians.

During this same time the Sioux signed yet another peace treaty with the U.S. government known as the Treaty of 1868. Both sides agreed to end hostilities and solemnly promised to keep the peace along the Bozeman Trail in Wyoming and Montana. Forts Kearny, Smith, and Reno were abandoned by the soldiers. The Indians agreed to turn over for discipline any braves found in violation of the treaty. The Bozeman Trail was declared closed to white traffic but the government retained the right to establish other forms of transportation—meaning new routes—as it deemed necessary. Provisions were made in the agreement for the customary annual handouts to the Indians as well as educational offers. The government also gave a tract of 160 acres to every adult brave along with implements to farm it. At last, the well-meaning proponents thought, the Indians could be taught to do an honest day's work contributing something worthwhile to society. It was a silly provision doomed to failure from the start. Native Americans had lived their entire existence as nomadic hunter-gatherers. They had no desire to become tied to any one piece of land as farmers. In their minds they were already contributing quite nicely to the society they lived in. They provided for their families by hunting and fishing. They captured and domesticated the horse as their main means of transportation. They raised their families and lived in harmony with nature. They were a happy people.

Article 2 of the Treaty of 1868 would come back to haunt the white man. It set aside most of the present state of South Dakota including the Black Hills as an Indian reservation. The article acknowledged this land as unceded territory. It was declared off-limits to both travelers and settlers. The unceded status meant that the land was an independent state completely and solely under Indian control. The frontier generals knew that sooner or later they would be called in to clean up the mess that this article created. Nevertheless, there was peace for the time being.

Close scrutiny of our relationship with the Native Americans is almost painful to reflect on. One cannot help feeling curious about Heath's opinion of these misunderstood people. Did he fear them after hearing tales of their savagery or even witnessing firsthand their butchery of his own kind? His early service with the Seventh Cavalry in the fall and winter of 1875–1876 gave him chances to see such atrocities. Might he have felt the hate toward them that is so often the direct result of ignorance and fear? Is it possible he cared in some way about our treatment of them? Perchance a twinge of sympathy or compassion was there when he thought about the inhumane way in which his country was dealing with its growing pains.

In spite of our present-day enlightenment, most citizens still view the Indian suspiciously as lazy and inherently containing a bad genetic seed. It is almost as if we whites need to justify the fact that we were the foreigners in this case; something we like to ascribe to the red men. Even the clemency-prone clerics of the missionaries, after observing the Indians' mythical tribal practices, declared them to be a species barely above an animal. The term they used was "savages." Kindly translated they meant uncivilized. In order to drive home this point they cited the practice of scalping. Had they done their homework they would have easily discovered scalping had started with the white man himself.

As early as 1755 a Massachusetts colony proclamation declared a bounty of forty pounds for every male Indian and twenty pounds for females or children under twelve years of age confirmed dead by their scalps. Vine Deloria reports in his work *Custer Died for Your Sins* that loyal colonists were urged "to embrace all opportunities of pursuing, captivating, killing, and destroying all of the aforesaid Indians."

The label of "uncivilized" was grossly unfair. These were a people who truly understood the meaning of caretaker and steward of the earth as mentioned in the white man's Bible. They took only what was necessary to survive and promptly gave thanks to the gods that the animal they killed allowed them to subsist. Indians never even conceived of the idea of individual land ownership or selling the trees covering the land or the rocks that lay buried under the ground for profit. Some would say their criminal justice system was far more advanced than ours. There were no prisons in Indian society. If a man wronged another the sentence was not retribution in the form of jail but compensation to the family that had been violated. These "savages" were highly religious dedicated family members and generally peace-loving nomads who roamed as free as the animals they followed. What was uncivilized was to herd them into human zoos called "reservations."

I will make one last reference to our concept of Indians as uncivilized. Chief Sitting Bull is probably the most recognized of the great chiefs to participate in the Indian Wars. Over the years his picture has appeared in history books and novels countless times. His weathered face is haunting as he looks at the camera through piercing eyes. With his hair in braids and a single feather in his hair he looks every bit the brave warrior that he was. Some years after the battle Sitting Bull became the star attraction in Buffalo Bill Cody's Wild West Show. While touring the major cities up and down the East Coast he made money. Historian Stephen Ambrose reports in his book *Crazy Horse and Custer* that Annie Oakley noticed he gave most of it away to the poor little children he found hanging around the show. Sitting Bull remarked that "The white man knows how to make everything, but he does not know how to distribute it." I'm certain Sitting Bull had never read the English philosophers who proposed the redistribution of wealth in our society, yet here was a man who practiced it every day of his life. Quite a feat for an "uncivilized savage."

The minute the white man and his government realized that the Indians were in possession of millions of acres of prime timberland, rich minerals, abundant animal life, and land suitable for agriculture and grazing, the intent to disenfranchise them was born. Initially the red men were given treaties allowing them free and undisturbed use of the lands; lands that were free and undisturbed until the arrival of the white man. Later, confined to reservations and sent to white man's schools, they were made to dress like whites, speak like whites, and worship like whites. This attempt to eradicate their heritage was the ultimate insult to a once-proud race.

The treaties started as a means for the white man to keep peace on the frontier. They rapidly evolved as a way to annex massive tracts of land. The very first time the Indians submitted to this practice marked the beginning of the end for them. None of these agreements were kept and I've never read a single history book that attempts to defend the idea that the government intended to keep them. The earliest treaties with the Indians and England were made to keep the Indians from joining forces with the French when the two were in competition for this country. How conveniently the English colonists forgot that it was the "savages" who helped the colicky infant America from falling prey to other nations. These were the same "savages" who helped the early colonists overcome exposure, drought, and starvation by teaching them how to survive in this harsh new land.

By the beginning of the twentieth century absolute power over the fate of the Indian was firmly implanted and culminated in the 1903 Supreme Court decision of *Lone Wolf* v. *Ethan A. Hitchcock.* Lone Wolf was a Kiowa Indian tribal leader who tried to get land allotted under treaty from the Interior Department. Hitchcock was the U.S. Secretary of Interior opposing this move. Ultimately the Supreme Court ruled that the Indians had no title to the land and never did. The Court said the land is and was owned by the U.S. government and that the Indians were no more than occupants. In essence they were little more than vagrant squatters and the government reserved the right to dictate any terms to them it saw fit.

In the year 1874 Ulysses S. Grant was president of the United States and William Heath was still living in Girardville, Pennsylvania. Both were starting to experience some rather tense developments in their lives. Heath's has already been chronicled. President Grant, being the former victorious general of the Union army, had the unwise habit of doling out political jobs as plums to former war buddies and cronies. Regrettably not all of them were deserving of their positions. One such appointee was Secretary of War William W. Belknap, who was corrupt.

General Sherman, in charge of the army, was so disgusted with Belknap and the rampant unprincipled conduct in Washington that he moved his headquarters to St. Louis. From here he communicated with Belknap only when absolutely necessary, using the telegraph rather than speaking with him personally. Charges were levied against the Department of the Interior regarding corruption at the Indian agencies and in the selling of land out west. Due to pressure for reform, Grant turned over the running of many of the Indian agencies to well-meaning religious groups. It was their idea to turn the Indians into farmers.

At this same time Sitting Bull's star as a leader had risen. His reputation for bravery and valor as a warrior was widespread. Many braves began to gravitate toward him; even those from other tribes. Some of them who had sworn their allegiance to Red Cloud had become disillusioned with his unexpected tolerance toward their old nemesis, the U.S. government.

Gen. U. S. Grant emerged from the Civil War as the most popular man of the day. Thankful citizens dug into their pockets to show their appreciation. In his hometown of Galena, Illinois, as well as in Philadelphia and Washington, he was given homes to live in. The city of New York presented him with a check for no less than $105,000. So popular was he in 1868 that he could have had the party nomination from either the Republicans or the Democrats, an unheard-of expression of endorsement.

Unfortunately, Grant was a simple-hearted, cigar-chomping soldier who was a better judge of horseflesh than he was of human beings. His two terms as president were to be stained by corruption and political naiveté. The Republicans shrewdly beat the Democrats to the punch by nominating Grant for the presidency in Chicago in 1868. The former general accepted the call of his nation by uttering the simple phrase "Let us have peace," which quickly became the slogan of his campaign. The Democrats, having lost out to the Republicans in obtaining the most popular candidate in the country, fell into disarray. Torn by factionalism, they struggled in the campaign with their nominee, Horatio Seymour, a popular six-time governor of New York. Grant easily bested his adversary by capturing 214 Electoral College votes to a mere eighty for Seymour.

The Civil War itself had spawned waste, graft, and corruption in government. Sadly, it both continued and flourished under the basically honest Grant whose major sin was blindness to the political ramifications of his decisions. As great economic and social changes developed, President Grant operated as if in a vacuum. Large numbers of judges and politicians could be had for the right price. It was often said by historians that during Grant's administration the definition of an honest politician was one who, when bought off, would remain loyal to the buyer. Unfortunately, the public looked upon these circumstances with indifference as outcries were few and far between.

Thus Grant's tenure was jokingly referred to as "The Era of Good Stealing." The president selected a cabinet that befuddled onlookers. Accomplished with the secrecy of a military covert action, not even his wife was privy to the details. When the cabinet was announced, it was obviously a profusion of Grant cronies, hangers-on, misfits, and greedy opportunists. The president faithfully rewarded those who hopped aboard the house-for-Grant train and the check-for-Grant train and remained loyal to them throughout his presidency. These dishonest friends and acquaintances led the unwitting president through a carnival of corruption that included civil service abuse, cruel and crooked reconstruction of the South, stock market scams, and scandals too numerous to mention. By the time U. S. Grant was being urged to take a third term in 1876 the public had seen enough. With a resounding bipartisan vote of 233 to 18 Congress passed a resolution, not a law, expressing its desire for Grant not to run again and the Republican Party, seeing the handwriting on the wall, did not nominate him for a third time.

Complicating Grant's presidency was a serious economic downturn. Businesses in the East were failing left and right. Unemployment rose to near-record levels mostly due to overspeculation in the gold market. Even railroad construction ground to a halt. Surely this must have brought a smile to the face of many a Native American. As the depression described earlier in Schuylkill County set in nationwide, intense pressure was brought to bear to locate new reserves of gold. Enterprising companies in the East sent prospectors to the Black Hills where rumors of gold had been seeping out for years. The government realized that a scientific expedition was needed to confirm or deny the rumors. Gen. George Custer was assigned the task.

The Black Hills expedition of 1874 was officially for the purpose of determining if the reports of gold were true. Unofficially, the army was also looking for an ideal spot for a military fort in preparation for opening up the land for speculation and settlement, something that had already been decided upon. The government knew full well that any cry of gold would set off a tidal wave of migration. The government also knew that once it started there simply was no way either Washington or the Indians could stop it.

Custer left for the Black Hills in June of 1874. With him were ten companies totaling about one thousand men. Although William Heath was not on this expedition with the general, astonishingly, another man from Girardville was. John Wallace Crawford, also known as Captain Jack the Poet Scout, rode directly alongside Custer in the expedition and was nearly as well known out west as Custer himself. Captain Jack was born in 1847 in County Donegal, Ireland. His father brought him to the United States when he was seven years old. According to the *Girardville Chronicles* by Lorraine Stanton, the family lived in Minersville while the father worked in nearby mines. At age seventeen he enlisted in the Union army and was wounded at Spottsylvania and again at Petersburg during the Civil War. A nun taught him to read and write while he recovered from his wounds.

After the war Crawford moved to Girardville where he was the borough postmaster in 1869. A short time later he headed west and became famous. At the time of the Sioux uprising in 1875 he served as a scout for Generals Wesley Merrit and George Crook when the position was vacated by Buffalo Bill Cody, a personal friend of the Captain. In addition, Jack Crawford ran a trading post in New Mexico, was an Indian agent for a while, and established a ranch on the Rio Grande. Sharing his time between the ranch and his home in Brooklyn, New York, the Captain published volumes of poems, wrote and produced three plays, authored over one hundred short stories, and gave numerous lectures.

The romp through the Black Hills was one of the most enjoyable experiences of Custer's life. The natural beauty coupled with the abundance of game amazed him. He reported that the hills were alive with fish, game, birds, deer, and elk in grassy valleys and on forested slopes. When General Custer returned from the Black Hills he reported finding gold. Predictably a flood of prospectors descended on the area, desecrating the Indian sacred ground. In a short time it was estimated that as many as fifteen thousand whites swarmed over the area. At first the army made a half-hearted effort to keep them out. Soon it looked the other way while President Grant worked on a plan to extinguish the Indians' title to the land, vowing to use all means at his disposal to accomplish that end.

Chapter Three

MOVING OUT

L ate in 1875 William Heath had already joined the army and was a member of the Seventh Cavalry. There was a growing sense of urgency among the Indians now. The white men were pouring into the Black Hills region, turning over every rock in their path in their search for gold. The Indians had seen this gold fever before in the white man and knew the consequences of it. The red man's sense of urgency was probably much like that Heath felt when he hastily left Girardville—a feeling of life and death.

Billy Heath must have looked good in his uniform. At five feet seven inches he probably weighed no more than one hundred fifty pounds (estimate of Dr. Richard Bindie, forensic pathologist, after examining Heath's photo). He was physically fit from toiling in the mines of Schuylkill County and needed little drilling with the Seventh to get into shape. The occupation of miner was handed down from his father who was listed as a colliery man in England according to family history. Now he had made the jump from miner to soldier.

Dressed in uniform with his brown hair and blue eyes he looked dashing. Sporting a blue wool blouse over gray shirt, he had on sky blue trousers tucked into a pair of military boots. Topping off his ensemble was his hat. The military issue of the day was of poor quality. It drooped immediately after getting wet and quickly began to fall apart. Most soldiers wore civilian-made hats (frontier style) for their durability and appearance. As a former coachman he was familiar with horses. His most recent stint with the Coal and Iron Police meant that the handling of firearms was not new to him. In retrospect Heath was more likely better prepared for battle than most of the other new recruits.

As William Heath was drilling and occasionally going out on missions with the Seventh Cavalry, things were beginning to heat up around him. General Crook was receiving increasing numbers of reports of Indian attacks on whites. The trail left by raids on people and stock led right to the Sioux reservation in the Black Hills. The government had a solution in mind even before Heath joined up. In September a delegation of Indians and representatives of the president met at the Red Cloud agency, an arm of the Department of the Interior to administer control over the Indians. The Presidential Commission wanted to buy the Black Hills, Powder River, and Bighorn areas from the red man.

The treaty of 1868 required a three-fourths majority vote by all Sioux males before it could be changed. Many Indians, when informed of the meeting, did not come. They chafed at being ordered to report. Eventually some twenty thousand did show up. The initial reaction of the chiefs was not favorable. But it was soon to get worse. Crazy Horse commented in disgust that one does not sell the earth upon which the people walk. Sitting Bull was just as poignant. When the letter of intent was handed to him he picked up a pinch of dust from the ground saying he would not sell even as much as that to the white man (see exhibit 14).

The red and white man sat together under a huge tent at the Indian agency. As the chiefs sat inside, thousands of agitated Indians milled about outside. Sometimes arguments broke out between the Sioux, the Cheyenne, and the Arapaho present there. An Indian guard circled the tent to protect the very nervous white men inside. For two weeks the haggling went on with neither side bargaining in good faith. The Indians never truly intended to part with their sacred ground. The white men knew full well that if the Indians didn't give it up the government would take it from them.

Exhibit 14. Chief Sitting Bull, circa 1880–1890. (Courtesy Denver Public Library, Western History Collection/Palmquist & Jurgens, St. Paul/X-31721)

Chief Red Cloud.
(*Dictionary of American Portraits*, Dover Publications, Inc., 1967. Courtesy Mercaldo Archives)

The commission tried offering the Indians more land beyond the setting sun if they gave up the Black Hills. Their next offer was for mineral rights of $400,000 a year for fifteen years or $6 million outright if the chiefs preferred. The Indians had been told by some well-meaning Christian Indian agents that their land was worth at least ten times that and were not impressed with the government overture. Finally Red Cloud spoke. He counteroffered with a request that the government feed and clothe all the Indians for seven generations plus give them $600 million and they would give up the Black Hills. By now many of the braves had already left with a bad taste in their mouths, vowing never to give up the land without a fight. The commission departed for Washington to report back to President Grant. Certainly, by now the president knew what his next course of action would be.

In early November 1875 Indian Inspector E. C. Watkins sent word to Washington that bands of Indians were making war on white settlers and each other, violating reservation rules. He asked for army intervention. According to the treaty it was the whites who were in violation, but that fact seemed to be overlooked. On December 3 the government ordered all Indians to report back to the reservation by January 31, 1876, or else. February 1 came and the Indians did not comply, prompting the secretary of the interior to notify the secretary of war that the Indians were now his problem. It should be noted that some of the Indians did come back to the reservation and Crazy Horse himself was reportedly on his way back in. Most, however, did not comply partially due to their attitude and partially because the brutal winter conditions were not conducive to uprooting villages and transporting women and children through the snow and cold.

Secretary of War William W. Belknap sent out instructions for the U.S. Army to take whatever measures appropriate to correct the situation. Commanding Field General Philip Sheridan was delighted with the order to move against the Indians. He had been openly frustrated by the Indian Bureau's policy on giving them a chance to come in on their own. Now he could do what a soldier had to do. He hastily formulated a plan for a winter campaign to whip the redskins into submission. Sheridan considered attacking at wintertime crucial to the plan's success. He knew the Indians would be immobile as they were holed up against the fierce weather. Also he knew they would least expect attacks from the white man during such conditions. In addition, wise in Indian ways, Sheridan was aware that the number of braves would be at its low point in winter. Once spring came, they traditionally gravitated together for the Sun Dance festival and then began following the herds of game. In winter many braves and their families stayed on the reservation where they could get food and shelter, meager though it might be, from the white man.

Sheridan's strategy envisioned a pincer-type attack on the Indians from three different directions. The two officers he put in charge of carrying out his plan were Gen. George Crook and Gen. Alfred H. Terry. One column under Crook was to come in from the south while two columns under Terry would converge from the north and east. According to intelligence reports from Indian agencies, the Indians were in the area of southeastern Montana a little north of the Bighorn Mountains. That is where the generals expected to have their showdown.

Information from Indian agents put the strength of the Indians at between five hundred to eight hundred braves. These numbers were fairly accurate. However, by the time hostilities took place in the summer the size of the Indian camp would grow to several times that amount. The generals never expected to find such a huge number and that fact alone contributed to the collapse of Custer. They always figured the Indians could never stay together in large numbers for any length of time since to do so would decimate the forage for their ponies, foul the water, and drive away life-sustaining game.

The crucial element of the plan, the winter offensive, never took place. The winter of 1875–76 was especially brutal, and by the time the word came to commence action the troops were penned in by the unrelenting ice

Gen. George Crook. (*Dictionary of American Portraits*, Dover Publications, Inc., 1967. Courtesy Library of Congress, Brady-Handy Collection)

and snow. The delay of this winter campaign also factored into the failure at Little Bighorn. The military, however, was never overly concerned with these facts. They figured their main problem would not be finding the Indians and defeating them. Their biggest fear was to keep them from escaping—another fatal error in judgment. This time the Indians would not run.

It wasn't until the spring of 1876 that the U.S. Army got its first taste of battle. On March 17, while searching for the enemy, Col. J. J. Reynolds under Crook's command attacked a peaceful village of Sioux and Cheyenne. Reynolds stormed in at dawn while the Indians slept. He burned their tepees and saddles, capturing over one thousand of their ponies. The very next night the surviving Indians stole back the horses and rode off to the northeast to join up with Chief Crazy Horse. Crook became so angry with Reynolds for allowing this that he later court-martialed him for his lack of diligence.

Crazy Horse immediately broke camp, moving north to the Tongue River. Soon after, Sitting Bull arrived with his people along with Chief Lame Deer and his followers. They settled in around the Rosebud Creek, close to the Little Bighorn River. Rumors flew in the now huge village about a large number of bluecoats coming with their old nemesis Yellow Hair (Custer). Still, the Indians did not believe that the soldiers would actually attack them. On June 17 the Indians took a small offensive action. A hunting party of Cheyenne spotted Crook's camp and reported their finding to Crazy Horse. A band of one thousand Sioux and Cheyenne attacked Crook, causing him to retreat in defeat. Called the Battle of the Rosebud, this action virtually knocked Crook out of the hostilities, thereby eliminating one third of the pincer movement General Sheridan had intended. Even though Crook did not suffer grave losses, he remained out of the action.

Just one month prior to Rosebud the nation embarked on the grand celebration of its first one hundred years. The U.S. centennial celebration opened in May at Fairmont Park in Philadelphia, Pennsylvania. There had been world's fairs before in the great cities of Vienna, Paris, and London but this was a first for the United States. Despite the pall of an economic depression it was a huge success both socially and economically. Over five months it drew nine million visitors; 275,000 in one day alone. There were displays from around the world: German porcelain, Japanese bronze, English painters, and even jewels from America's own Native Americans.

The main building covered an amazing twenty acres, making it the largest building in the world as it spread out along the Schuylkill River. The centennial party ignited a wave of patriotism for what we had accomplished in a brief one hundred years. Optimism, which had been lacking due to the depression, reemerged. Newspapers editorialized that if so many people could afford to travel to Philadelphia things simply couldn't be all that bad. Our confidence grew as the populace began to realize that this country was a diamond in the rough just waiting for us to cut it to perfection.

❧   ❧   ❧

From the picture painted thus far it is not difficult to perceive that attacks from the United States Army were merely the last straw that preceded the Battle of Little Bighorn. For decade after decade actions perpetrated by white invaders were destroying Indian lands and their very way of life. Trappers crisscrossed the land seeking furs for profit (something the American Indian couldn't comprehend). Settlers appeared in Conestoga wagons and cleared the land for farming. Cattlemen closed off the land to graze their great herds of cows and beef cattle. Railroads sliced into the land, decimating the buffalo herds as they went. Evangelists followed to convert the savages to Christianity, denying them their own religious beliefs. Finally, there was gold found in the Black Hills.

It is no great leap then to understand the frustration and anger the red men felt toward the whites. They were losing their land to fences and farms. Their game was disappearing and their sacred ground being raped to satisfy the white man's lust for gold. Finally, they were being herded onto reservations and denied the freedom to roam the land as they did for generations. Left with little choice they fought back at Little Bighorn.

Exactly what happened to William Heath and, in particular, the Seventh Cavalry that hot day in June is the subject of immense controversy. Many pages have been written, and much of what has been penned is steeped in speculation. Since the claim has always been made that all soldiers who were with Custer's battalion were killed, there was no one left to tell the true story. It has been pieced together by historians, archeologists, and forensic scientists. Their hypotheses are both many and varied. In studying the various works on the battle in preparation for this volume I found each one a little different: different time lines, different distances, different numbers, and different places to put the blame.

However, in reality there were hundreds of participants who were there and who survived. The Indians who took part have given many credible accounts, but the white man has largely either ignored or discredited them. This is an unfortunate reflection on the white man's character since the Native Americans of the nineteenth century were generally regarded as an honest and forthright group of people. I do not entertain any thoughts of trying to prove one particular theory of that day's events. What I have attempted to do is to include as many accounts of it as possible so that the full range of possibilities is explored. A list of all sources can be found in the bibliography. Regardless, none of the theories on the battle in any way destroys the basic premise of this book, the fact that at least one soldier did survive.

🌿   🌿   🌿

May 17, 1876. Fort Lincoln sits strategically on the Missouri River near the Northern Pacific Railroad in the Dakota Territory. There are people inside lined up with anticipation. There's a parade today. The parade grounds occupy center stage with various buildings containing mess halls, barracks, offices, and the like arranged in orderly military fashion around it. The civilians are somewhat agitated, as they have been waiting for this event for days. A three-day deluge of rain that turned the entire valley around the fort into a muddy mess has delayed the departure of the Seventh Cavalry. Now they are ready to witness the army's finest display before they march off to battle the enemy.

William Heath sat on his bay horse in the middle of one of the four columns of Company L. He had on his canvas-lined pants to protect him from the weather and from saddle sores. A saddlebag was tied to the saddle along with one blanket and an overcoat. Hanging from his pommel was everything needed for the care of his mount: nosebag, lariat, and stake for picketing. As farrier for Company L, it makes sense that Heath had selected for himself one of the best horses in the bunch: strong, sure-footed, and calm under fire. Making such a selection could and would be a matter of life or death.

Fort Abraham Lincoln, winter. (Courtesy Denver Public Library, Western History Collection/D. F. Barry/B-838)

All William's personal necessities were carried in his haversack. It contained his tin cup for drinking coffee, a poncho for bad weather, his mess kit for eating, some matches, spare ammo, a twist of tobacco, some salted pork, and a tin of crackers. Strapped to his side was an 1873 model Colt .45 revolver. In his saddle pouch rested his single-shot breech-loaded .45 caliber Springfield rifle. Although there were newer repeating rifles available, the army chose this one for its superior hitting force and better accuracy over long distances. Neither Heath nor others of the Seventh took along sabers. They were used only in hand-to-hand combat, and the Seventh didn't see the necessity to bring them. If you found yourself that close to the Indians you were probably in big trouble anyway.

Young Billy Heath could not help but be moved by the wave of patriotism that swept over the crowd as he marched past. His company was one of twelve that paraded for review. There was a full brigade band, mounted on white horses, playing the Seventh's favorite song, "Garry Owen." This band was another example of Custer's flamboyance. Company A rode blacks. Companies D, F, H, I, L, and M rode bays. C, G, and K had sorrels with E on grays and Company B, which guarded the supply wagons, was the only one with mixed colors. The twelve companies painted quite the picture marching in review.

Each company, when at full strength, was to have one hundred privates, corporals, and sergeants plus a captain and two lieutenants. As Heath and the Seventh marched on parade this overcast day they stood at roughly six hundred fighting men or about half their authorized full strength. After marching around the parade grounds they broke ranks for one last good-bye to friends and family. Billy Heath had no family to wish him luck. His wife, Margaret, and their two sons were back in Pennsylvania unaware of the historical adventure he was about to undertake. He probably spent the time making sure he had everything he would need for the campaign; especially the one hundred cartridges for his Springfield rifle and twenty-four for the Colt revolver. If only Margaret and the children could see him now.

When the roughly one-thousand-man contingent of the Dakota column under the command of General Terry finally left the fort, the band softly played "The Girl I Left Behind Me." The column that left Fort Laramie was impressive by the standards of the day. Joining the one thousand men and officers was a supply train of one hundred fifty wagons pulled by mules carrying food, ammunition, and hay. Trailing the wagons was a herd of spare horses as well as a herd of cows for food. There was a small detachment of the Twentieth Infantry along with their three rapid-firing Gatling guns, plus men from companies of the Seventeenth and Sixth Infantries. Rounding out the group were between twenty-five and thirty-five Arikara scouts as well as six Crow scouts. Lt. Charles Varnum, fresh out of West Point, was in command of the scouts. Two plainsmen also rode with the troops. Mitch Bouyer was half-Indian and knew every Indian camp in the territory. Lonesome Charley Reynolds, a savvy veteran of Indian ways, held the reins of his horse in one hand. His other hand (including his trigger finger) was confined to a sling because of a nasty infection.

The Seventh Cavalry had many good men with them, men William Heath knew well and had come to count on. They had at least three West Point graduates, men from the Foreign Legion, the Queens Guard of Canada, the Papal Guards from Italy, and even a trumpeter who had been a drummer for Garibaldi himself. Some officers like Gibbs and Smith were experienced Indian fighters. Lt. Ed Mathey came all the way from France to fight for the army. Captain Meyers was a German immigrant. Lts. Jackson and Nowlan were experienced soldiers from the British army. The oldest officer was fifty-four-year-old Capt. William Thompson. The youngest had been twenty-two-year-old Capt. Louis Hamilton, the grandson of Alexander Hamilton, who died earlier at the Battle of the Washita.

According to Peter Panzeri in his book *Little Bighorn 1876*, the officers of the Seventh were strange bedfellows by anyone's standards. Panzeri says they were "one of the most petty and bizarre command climates on record." His research, as well as that of others, gives more than adequate fodder for his charges of "alcoholism, cowardice, favoritism, and questionable fraternization between officers [and enlisted men]." Although Custer is believed to have kept strict control over his men, there was still much conflict in the unit. Much more than prudent in a well-oiled fighting machine about to do battle.

The Seventh Cavalry would operate according to the 1874 Army Manual of U.S. Cavalry tactics written by Emory Upton. It was not a system well suited to fighting Indians but more for Civil War tactics. Battalions consisted of two to seven companies. A company, when at full strength, had sixty men in it. Custer's companies contained anywhere from thirty-eight to forty-four men. Platoons were subdivisions of a half of a company while a squad was made up of four mounted soldiers. Each company was organized by the assignment of command going to the senior ranking officer in it. Battalions would be subdivided into wings (left and right) as the commander saw fit. Being mounted the Seventh Cavalry was a mobile force; a necessity when pursuing fleeing Indians. However, the rigid organization and blind adherence to proper tactics robbed them of all flexibility and was a serious defect. One can't help but wonder if William Heath took notice of this defect as he served with the Seventh during the winter of 1875–1876 preceding Little Bighorn. If so his concerns would have fallen on deaf ears. Also, it is interesting to guess where Heath may have fit into the social strata of the Seventh. Was he an active, boisterous member of Company L or did he keep to himself, avoiding any unnecessary contact with this unusual group of officer gentlemen and enlisted men of questionable lineage—men on the run from the law or those lacking education? Going by what we know of Heath's demeanor, he probably was a loner. Most likely he participated sparingly in the daily life of a cavalry trooper, just enough to gain acceptance while always guarding his private thoughts.

As the soldiers and Custer (dressed in a light buckskin coat and frontier hat) moved out of sight, one of the last things to be seen was the red scarves worn around the necks of the Seventh's troopers, another of General Custer's affectations designed to set the Seventh apart in a class by itself. Libbie Custer, the general's wife, remembered later going into the house after the column disappeared from sight with an eerie sense of doom pervading her being, something she had never felt before. The Dakota column had not gone very far when they turned west into the blazing sun for the Yellowstone. Sheridan's winter war had now melted down in a hot summer search-and-destroy mission.

For weeks prior to Terry's moving out on the seventeenth he had put men out in the field trying to get an exact fix on the enemy. Col. John Gibbon with a detachment of the Montana column was moving into position to the north of Terry's intended line of march to prevent escape should the Indians flee in that direction. Strangely, he had located the Indian camp on the Rosebud as early as the end of May and never conveyed that information to Terry, who was vigorously searching for this target.

It wasn't until June 8 that Terry met up with the Montana column and learned the village's location. General Terry wanted to make sure the Indians did not double back, so he sent Major Reno on a two-hundred-mile southward swing to check the Powder and Tongue River valleys. Reno, in violation of orders that directed that he not go beyond this point, crossed as far as the Rosebud to make sure of his mission. Here he found recent campsites and returned to Terry. By now the Indians were no longer camped where the Montana column had seen them some two weeks earlier. They had relocated from the Rosebud to the valley of Little Bighorn. When Reno returned with his report he felt the Indians were no more than a mile or two from his last position. The truth was they were more like sixty miles away. This lack of accurate knowledge of the enemy's whereabouts necessitated a longer march in depleting the energy, as well as the resolve, of both men and horses. Thus it played a significant role in the outcome of the battle.

General Terry boarded the steamer *Far West* that was there to transport supplies up the Yellowstone River. He had Custer march the rest of the

Steamboat *Far West*. (Courtesy Denver Public Library,
Western History Collection/D. F. Barry/B-798)

troops overland to meet with him in about two days' time. General Custer
linked up with Reno and the other half of the Seventh and together they
marched toward the designated meeting place along the river.

At noon on June 21 General Terry met with Custer and Gibbon in his
office on the *Far West* and revealed a revised plan. Information from Reno
and the Montana column forced him to change his original plan of
keeping Gibbon on the Yellowstone to the north. Instead, Terry sent
Gibbon south on a parallel track with Custer. The idea was to squeeze the
Indians between the two contingents. Custer was to follow the Little
Bighorn north while Gibbon would traverse down the Bighorn to the Little
Bighorn where the two would link up. Once one of them engaged the
enemy the other would theoretically be in position to reinforce. General
Terry decided to march with Colonel Gibbon. Everyone was ordered to
take an ample supply of salt in case the march extended beyond their fif-
teen-day supply of rations, forcing them to live off live game.

While the commander of the Seventh Cavalry met on board the *Far*

West, its soldiers rested on land, perhaps chatting about where they were headed next. Billy Heath might have moved along the ranks looking for two men he probably knew by now, having found something in common with them. One was right in his own Company L. The other could be found in Company F. Against unbelievable odds, there were two other men besides Heath from Schuylkill County about to go down in history with him. George Adams of L Company was born in Minersville in 1846. His military record suggests he was somewhat of a colorful character, one with whom Heath might have enjoyed conversing.

Adams enlisted in the United States Infantry on October 18, 1869, at Fort Randall in the Dakota Territory. His previous occupation was listed as a teamster (which meant he may have driven wagons out west as a civilian). He was court-martialed by the army in 1870 with no reason given. One week before Christmas in 1872 he was discharged from service for reason of disability (no specifics given). Adams reenlisted with the Seventh Cavalry on January 27, 1872, at age twenty-seven and a half years at Fort Lincoln, listing his previous occupation as soldier. One year later, on January 3, 1875, he went AWOL, returning to his unit a second time since being discharged, but was acquitted in March. By April he was in trouble again. On this occasion he was charged with being drunk on duty the previous month. Court-martialed for a third time in May he was sentenced to six months' hard labor for this violation. Following this last incident he appeared to have straightened out and served without any further problems.

The second Schuylkill County resident in the Seventh was Pvt. Herman (sometimes listed as Harman) Knauth from Brandonville, Pennsylvania (a small town in the northern part of the county). Knauth was born in Dammendorf, Prussia. At the age of thirty-three he enlisted in the army on January 20, 1872, in Rochester, New York. His occupation was listed as merchant. He was a Heath look-alike with blue eyes, brown hair, and standing only one-half inch shorter than Heath at five feet six and one-half inches tall. One could easily mistake the body of the one man for the other on a bloody battlefield. Nothing further is known about him.

While the men of the Seventh sat talking and waiting for their leader to return, a sudden violent hailstorm erupted. In a matter of minutes the entire ground around the *Far West* was turned white with hailstones. The men grabbed their frightened horses and sought cover wherever they could

find it, since some of the stones were large enough to do damage. Soon
enough Custer came out and told them to bed down for the night. They
were told they were moving out in the morning.

Keeping in mind that General Terry did not yet know of the defeat
handed to General Crook at the Rosebud, official written orders were
delivered to General Custer the next morning. The controversy over the
wording of these orders is so hotly debated that I reproduce the orders
here in their entirety.

> Lieut. Col. G. A. Custer, 7th Cavalry
> Colonel:
>
> The Brigadier-General commanding directs that, as soon as your
> regiment can be ready for the march, you will proceed up the Rosebud
> in pursuit of the Indians whose trail was discovered by Major Reno a few
> days since. It is, of course, impossible to give you any definite instruc-
> tions in regard to this movement, and were it not impossible to do so, the
> Department Commander places too much confidence in your zeal,
> energy, and ability to wish to impose upon you precise orders which
> might hamper your action when nearly in contact with the enemy. He
> will, however, indicate to you his own views of what your action should
> be, and he desires that you conform to them unless you shall see sufficient
> reason for departing from them. He thinks that you should proceed up
> the Rosebud until you ascertain definitely the direction in which the trail
> above spoken of leads. Should it be found (as it appears almost certain
> that it will be found) to turn towards the Little Horn, he thinks that you
> should still proceed southward, perhaps as far as headwaters of the
> Tongue [about twenty miles south of the Rosebud], and then turn
> [northwest] towards the Little Horn, feeling constantly, however, to your
> left, so as to preclude the possibility of the escape of the Indians to the
> south or southeast by passing around your left flank.
>
> The column of Colonel Gibbon is now in motion for the mouth of
> the Bighorn. As soon as it reaches that point it will cross the Yellowstone
> and move up at least as far as the forks of the Big and Little Horns. Of
> course its further movement must be controlled by circumstances as they
> arise, but it is hoped that the Indians, if upon the Little Horn, may be so
> nearly enclosed by the two columns that their escape will be impossible.
> The Department Commander desires that on your way up the Rosebud
> you should thoroughly examine the upper part of Tullock's Creek, and
> that you should endeavor to send a scout through to Colonel Gibbon's

column, with information of the results of your examination. The lower part of the creek will be examined by a detachment from Colonel Gibbon's command.

The supply steamer will be pushed up the Bighorn as far as the forks if the river is found to be navigable for that distance, and the Department Commander, who will accompany the column of Colonel Gibbon, desires you to report to him there not later than the expiration of the time for which your troops are rationed, unless in the meantime you receive further orders.

Very Respectfully,
Your Obedient Servant,
Ed. W. Smith, Captain, 18th Infantry
Acting Assistant Adjutant General*

The key wording to the supporters of Custer's actions at the Battle of Little Bighorn is that General Terry did "not wish to impose upon you precise orders." It is generally viewed as giving Custer a free hand to do as he thought best. Other Custer scholars contend just as convincingly that General Terry's words of his "own views" and "desires that you should conform" are clear indications of what Custer was expected to do. When a commanding officer expresses his desires, any military man knows if he does not heed that advice he runs the risk of incurring the wrath of his commander.

From what we know, Terry's plan distinctly intended for Custer to report his findings to him before engaging the enemy. Also its intent was for no offensive action to take place until everyone was in position to prevent the Indians from escaping. In both respects it appears as though Custer went against the better wishes of his commanding officer. Custer also made what many feel was a mistake when he foolishly declined the offer for the contingent of three Gatling guns to accompany him. His reason was that they would only slow him down. Their presence, as it turned out, could have made the difference in the outcome.

In a strange twist of fate the dashing boy general who led William Heath off to battle almost missed being there himself. George Armstrong Custer had been suspended from his command of the Seventh Cavalry and got it back only weeks before the battle. Back east there were charges

*From Charles M. Robinson, *A Good Year to Die* (New York: Random House, 1995).

of corruption in the dealings of public officials out west. Dishonesty in the granting of licenses (that amounted to monopolies) to run trading posts at military forts was charged, as were questionable practices in the sale of land and the malfeasance of many Indian agents on the reservations. The scandal involved several cabinet members and even led back to President Grant himself through his younger brother Orvil.

Custer was called back to Washington to testify before a congressional committee looking into the charges. Through Custer's deposition Orvil Grant was implicated, causing the angry president to suspend Custer in retaliation. Custer fought back by going to the newspapers to attack the administration, further outraging the president. The hearings took place in late March and early April and the president got great pleasure in seeing Custer squirm as he languished in Washington while the Seventh Cavalry was getting ready to do battle.

On three separate occasions Custer went to the White House to personally appeal to the president to reconsider. The first two times he was quickly turned away. The third time he sat for five hours, leading him to believe he was about to get an audience with Grant to plead his case. In the end he was rebuffed and never got to speak with the president. Custer broke down and appealed to General Terry in tears. Terry put in a plea to General Sherman who personally went to the president on Custer's behalf. Finally on May 8 Custer was reinstated. He couldn't leave Washington fast enough to get back west.

Gen. George Armstrong Custer, in my opinion a man by this time obsessed with political ambitions, never seemed to grasp the concept that his future was subject to forces beyond his control. He steadfastly clung to the belief that Custer luck could see him through any ordeal. He held onto an unrealistic self-image and fancied himself with facilities and skills beyond what he truly possessed. He failed to realize that his success was not so much due to his ferocious fervor in political and military exploits as it was to the favors of his political and military superiors. Perhaps he got a brief glimpse of this during his eye-opening incident with President Grant in Washington. Still, almost predictably, by the time he had returned to the Seventh he had the same old mindset: pull off an astounding military victory against the Indians and all was his for the taking.

It was during this same time that Gen. William Tecumseh Sherman

Gen. Alfred Howe Terry. (*Dictionary of American Portraits*, Dover Publications, 1967.
Courtesy National Archives, Brady Collection)

was embroiled in a bit of controversy of his own. Apparently he had undertaken a scandalous relationship with a beguiling young beauty by the name of Vinnie Ream. The Washington press dubbed her the Prairie Princess (she was born in Wisconsin) and it seems anyone who was anyone in the Capitol knew who she was. Vinnie was a striking beauty with sparkling brown eyes and tresses of thick curly hair who at age sixteen had charmed her way into studying as a pupil of sculptor Clark Mills in the basement of the White House. Soon politicians found their way to the basement to have a look and within a year she had done busts of Rep. Thaddeus Stevens of Pennsylvania and Sen. James W. Nesmith of Oregon.

Age nineteen and quite the lobbyist by now, she won the commission to do a full-length statue of Abraham Lincoln in part by a petition signed by then-president Andrew Johnson as well as his cabinet, thirty-one senators, and thirty-one other notables including Gen. George Custer. Some said the self-portrait bust, which depicted herself naked to the waist, didn't hurt either. A few years later she was linked to General Sherman. The fifty-three-year-old commander was smitten and wrote letters to the twenty-six-year-old beauty that suggest intimacy. He cautioned her to destroy his letters—which she didn't—and they are in the Library of Congress among her papers. When she married a young army lieutenant in 1878 in a wedding that was the social highlight of the Capitol that year, it was General Sherman who reluctantly gave the bride away.*

It is interesting to note that instructions for Custer's reinstatement included orders for Custer to be prudent and not to take along any newspaper men; both of which the young general violated at his first opportunity. Walking down the street hours after being reinstated, Custer bumped into a Captain Ludlow. Custer promptly told Ludlow he was going to cut loose from General Terry (the very man who interceded on his behalf) at the first available opportunity. When Custer pulled out of Fort Lincoln, Mark Kellog, a newspaper correspondent, was part of his entourage.

At the time General Custer himself was under suspicion of wrongdoing. Frederick Benteen (Custer's old enemy) had leveled charges of dishonesty in Custer's dealing with trading post officials. Benteen claimed Custer took a kickback of $1,100 from the head supplier of the famed

---

*For an interesting article with great insight on the life of Vinnie Ream, see Kathryn Jacob, "Vinnie Ream," *Smithsonian* (August 2000): 104–15.

Black Hills expedition. He further accused Custer of keeping his men off limits from any trading post who would not pay him off. Benteen had supporting evidence from some other officers as well. Custer was also linked by letter to the Quartermaster General of the Army Colonel Rufus Ingalls in a plot to get the army to purchase a certain brand of horseshoes for all the army's mounts. There was talk of Custer also being involved in a shady stock deal on Wall Street as well. Such were the circumstances under which G. A. Custer raced back west in May of 1876.

How William Heath viewed General Custer is not known. Heath had not been in the army a very long time and some of that time Custer himself was not present, so it is a difficult topic to address. As has been written before, some would follow Custer into hell while others would prefer to go over the hill rather than serve under him. If Heath were the deserting kind (and I bring this up now since critics are sure to bring it up later) he would have done so by the time of the Battle of Little Bighorn. We have already seen that the desertion rate for the Seventh Cavalry was a high 10 percent. Most deserters were brand-new recruits waiting for nearness to a gold find or a well-traveled trail, and not the least bit interested in the army's mission. Heath was none of these. He had suffered through a bitter winter with the Seventh and would have been gone at the first sign of spring if that were his inclination. He did not come west for the best of reasons but neither did he come to rush to the gold fields at first opportunity. He was not the career army type, yet he must have had a feeling of patriotism for in his last years of life he told his daughter Lavina that he was immensely proud of having served with the Seventh Cavalry. I suspect that out in the West, like back home in the East, William Heath was a company man. He followed orders and did his best to conform to his superiors' wishes. He would stay the course.

America has had a torrid love affair with the Wild West ever since the dime-store novels of the 1860s. The preeminent author of these novels at that time was Ned Buntline. He penned some of them before ever venturing west of the Mississippi. People like Jesse James, Judge Roy Bean, the Earp brothers, Sam Bass, Billy the Kid, and Bat Masterson are names known to most of us through these paperback novels. There's something about a rider on horseback, fearless and adventuresome, that captures our romantic imagination. The saga of Gen. George Armstrong Custer is one legend that towers over them all.

William F. "Buffalo Bill" Cody. (Courtesy Denver Public Library, MSS Collection Cody, William Frederick, Scrapbook vol. 1/X-22142)

Buffalo Bill Cody was perhaps the personification of the hero of the Buntline novels of the Wild West era. Born William Frederick Cody on February 26, 1848, in Scott County, Iowa, he moved to Kansas when he was eight years old. By the time he was eleven he had embarked on his famous career as Champion of the West. His first job was that of messenger boy for a freight company. Feeling the lure of the West, he traveled with various wagon trains, tending to the livestock and driving wagons. In 1860 he did a brief stint as a rider for the now famous and short-lived Pony Express delivering mail. When the Civil War broke out he served as a scout for the Union army.

After the war his adventures earned him his niche in American history. Bill Cody took up scouting and buffalo hunting for the railroads as they cut across the plains. His flawless marksmanship quickly earned him the nickname Buffalo Bill. During the years 1868 to 1872 he served as a civilian scout with the military, and for his bravery during a battle with the Indians at the Platte River he earned the Congressional Medal of Honor. Strangely, the medal was revoked in 1917 when Congress realized that he was not a member of the army when it was awarded.

Ned Buntline embellished the legend of Buffalo Bill so much that Cody himself was amused by the tales of his exploits. When the two finally met in 1869 they became fast friends and a new series of books by Buntline further enhanced the legend. By 1872 Buffalo Bill had begun appearing in theater productions of the Wild West. Soon he began taking his traveling show throughout the United States and Europe where he played before royalty. The show reenacted famous Indian battles and featured impressive stunts, trick riding, and feats of marksmanship. When news of Custer's defeat at Little Bighorn reached Cody he shut down the show and volunteered to serve the army as scout.

On July 17, 1876, at War Bonnet Creek Cody got revenge for his personal friend Custer and the Seventh Cavalry. There are conflicting stories as to the details but Cody appears to have done battle with the Cheyenne warrior Yellow Hair (which was the same Indian name given to Custer) whom he killed and then scalped in the name of Custer in full view of the rest of the braves. Cody returned to the stage with his popular appeal now greater than before as he played before full house after full house. During each show he proudly displayed the scalp of Yellow Hair to the enthusiastic audience. Buffalo Bill Cody died in 1917 on his ranch in the Bighorn Basin of northwestern Wyoming.

By the time the Seventh marched away from the *Far West* steamer on its seek-and-destroy mission Billy Heath was a firmly entrenched member of the group. His foreign roots and unusual motivation for joining up fit right in with the odd mixture that comprised the Seventh Cavalry and has been alluded to. Ranging from officers of distinction to enlisted men of questionable character, they shared the common goal to wipe out the red menace. Heath was almost a veteran now, having survived the brutal winter as a trooper. He had made a few friends, learned his role as a farrier, and was as ready as he ever would be to ride into history with the Seventh Cavalry.

Chapter Four

A GOOD RUN RUINED

E xcept for the present-day infatuation with the life of President John F. Kennedy, seldom has a man's life been looked into in more detail than that of General Custer. Relegated to true American hero status, a short biography of the man who led Billy Heath and the Seventh Cavalry into battle is appropriate. The bold and fearless slayer of savages was born on December 5, 1839, in New Rumley, Ohio. He was the product of the union of a widower and a widow, Emanuel Henry Custer and Maria Ward Kirkpatrick, who both had children of their own; a kind of modern-day version of hers, mine, and ours. At the age of fourteen he moved to Michigan to live with his half-sister Ann Reed with whom he was very close. She was fourteen years his senior. Working hard on her farm he developed great endurance for physical feats. Rather large for his age, George was always full of energy and constantly playing pranks on his friends and family. In 1855 he returned to Ohio and finished school. Records indicate that he was a marginal student at best, a reputation that would follow him to West Point.

As a young man he pursued young women with a reckless abandon and an unconventional boldness for the times. By age twenty he had written accounts of bedding down at least four separate lasses. Attracted by the twenty-eight dollars a month for five years Custer sought and acquired an appointment to West Point in 1857. During his tenure there he was notoriously undisciplined and earned demerits like compound interest. One hundred demerits in six months meant dismissal from the academy. George amassed 726 in four years' time. Academically he was constantly on the brink of being ejected. However, in all the physical requirements of the Point he stood out. He could best anybody in saber competition, calisthenics, and horsemanship.

During his last year he faced academic dismissal. If he did not pass a rigorous reexamination, he was out. A desperate Custer broke into the instructor's desk to steal the questions. As he was copying them he heard footsteps and, panicking, he tore the pages from the book and ran. In spite of the questions being changed when the tampering was discovered Custer was the only one out of thirty-three cadets retested to pass. Thus began the famous legend of Custer's luck. Hereafter he seemed to move with the full confidence of always finding a way out of a sticky situation.

He frequently stole away from the academy to go drinking and gambling. The former he managed to walk away from, the latter remained one of his vices despite promising his wife he would stop. Before graduating he gained the distinction of being court-martialed for, as officer of the guard, not breaking up a fight between two cadets. Custer's luck prevailed and he ended up with only a reprimand. Custer graduated in 1861 just as the Civil War broke out with the dubious honor of being last in his class. He swore his allegiance to the Union and entered the war.

Not long after joining the Union army Custer managed to get assigned as Gen. George B. McClellan's aide with the rank of captain. He served with McClellan at Antietam and gained instant distinction for rallying his troops and keeping a cool head in the heat of battle. It was in 1863 while on leave at his sister's home in Michigan that he first met Elizabeth Bacon. She was a stunning beauty with legendary chestnut hair and complementary figure. Bacon's father, a much-respected judge, was not fond of Custer and even Elizabeth was cool to him at first. Despite his show of affection for her, Custer still managed to find the time to pursue other women.

The year 1863 proved to be good for Custer. In June he received a field

Elizabeth Bacon "Libbie" Custer. (Courtesy Denver Public Library, Western History Collection/D. F. Barry/B-942)

promotion and became, at age twenty-three, the youngest general in the Union army. He was on a roll now as his luck continued. Two weeks later at Gettysburg General Custer, while commanding the Michigan Brigade, met and defeated the South's Jeb Stuart and his gray horsemen. Repeatedly, in campaigns through Pennsylvania and Virginia, Custer rode to victory time after time. By now he had developed the Custer look: blue shirt, black velvet jacket, spurs, red necktie, stars on his shoulders, long curls of blond hair cascading down, his blue eyes, and a ruddy complexion. He took a full brigade band everywhere he went.

The "Custer look." (Courtesy *Dictionary of American Portraits*, Dover Publications, Inc., 1967)

Custer returned on leave to Michigan again in late 1863 and set his sights on Elizabeth Bacon. He swore to her that he would give up drinking and gambling if she would have him. He told her she would have to live with the profanity, however, as it was a necessity on the battlefield. Libbie (as she was known to all who were dear to her) was concerned about the cussing and the fact that George never attended church. She wanted a year to think about it. George pressed her and by February of 1864 she gave herself in matrimony to him. From then on, whenever possible, wherever George went on assignment Libbie went with him. If being at his side was impossible they exchanged daily letters professing their love for each other. Custer scholars report that the letters reveal George and Libbie had a rather intense and diversified sexual relationship together.

In April of 1865 when Gen. Robert E. Lee surrendered to Gen. Ulysses S. Grant he did so by presenting himself to General Custer. At age twenty-five Custer's record during the Civil War was unblemished. In battle after battle he saw opportunity and took full advantage of it. He was now recognized as a true leader of men. The end of the war signaled a change for the boy general. When being mustered out of the Union army his rank was reduced to captain and later raised to lieutenant colonel. After the war, rank reverted to what it was prior to field promotion. He ended up with frontier duty in places like Texas and Kentucky, inspecting horses for the army. He went to New York to explore a career as a tycoon on Wall Street. He tried his hand at becoming a writer. In 1866 Custer was put in command of the Seventh Cavalry and within six weeks ninety men went over the hill. The men portrayed him as a tyrant displaying outright cruelty to them and gross disrespect to his officers. Custer became depressed and very moody. He was confused and definitely not showing the leadership qualities of earlier years.

Custer's luck shown through briefly when in 1866 he got his rank of general back by brevet.* No sooner did this happen than Custer made a move that almost destroyed his military career. In July of 1866 the erratic Custer force-marched some seventy-six men 150 miles in fifty-five hours to get him to a train to meet his beloved Libbie. As if that was not bad enough, on this same trip he left two of his men behind when they were

---

*Which means honorary (not unlike an honorary doctorate) and is why I refer to him as "General" throughout this book.

attacked by Indians; he was not on leave at the time and so had no business meeting his wife; and he had used government property—a wagon to transport Libbie and the men—for his own personal use.

A court-martial ensued and George Armstrong Custer was found guilty and was suspended without pay for a period of one year. He was on a roller coaster; one week he was up, the next week he was guilty of some unstable, almost abnormal behavior. During his suspension, Custer often suffered from bouts of depression. Custer tried defending himself in newspaper stories. By using untruths about dates and certain facts to suit himself he only made things look worse. But Custer's luck would come to his rescue once again. After serving ten months of the one-year suspension Custer was called back to duty. An Indian uprising required his presence, and he was reactivated ahead of schedule. The boy general needed something special to reestablish his tarnished reputation, something to prove that everyone had been wrong about him. That something turned out to be the Battle of Washita; the gruesome details of which have already been related in chapter 2. Custer scholars often use the troubles that Custer was having at the time as an explanation for his behavior at Washita. But I have another theory, one that I will elaborate on in chapter 5.

General Terry's plan of attack, although seemingly sound, lacked one very important piece of information: the size of his enemy. In the days prior to Little Bighorn the size of the Indian village grew to more than twice its original number, from about four hundred lodges to almost one thousand. The number of braves rose from eight hundred to close to two thousand. There were three reasons that this occurred. First, the Indians traditionally left the reservation by springtime to start following the herds of game. Second, it was time for the annual Sun Dance festival; an event the Indians looked forward to each year. Last, there was word among the Indians that a big fight was coming led by none other than Yellow Hair (Custer) himself. The Indian forces rallied around their many great chiefs.

At a little past noon on Thursday, June 22, 1876, William Heath and the Seventh Cavalry marched away from the steamer *Far West*. If Heath were close enough at the time he would have overheard Col. John Gibbon say to the young general sitting astride his faithful horse Vic, "Now, Custer, don't be greedy but wait for us." "No, I will not," stammered Custer in an unusually brief answer. Custer trotted past William Heath and the rest of

Company L to take his rightful place at the head of the column. He was now cutting loose from Terry as he had promised he would.

As the reality set in that they were going to battle, the troops became more serious as they rode on. The general himself turned noticeably moody. Pvt. John Burkman, who sometimes acted as Custer's servant, was abruptly ordered to Custer's side. Custer demanded to know why he was riding with the mule train instead of with his company. The man replied that he was doing so under the orders of the officer of the day. Custer sent him away and shortly thereafter rode back to apologize to the man.

The first day out the Seventh Cavalry made between twelve and thirteen miles before stopping for the night. As early as that first night in camp Custer's behavior was uncharacteristic. As Heath lay resting on his blanket after the evening meal, the officers passed by him on the way to a meeting called by the general. He told his officers to be careful to conserve rations as they could be out in the field longer than expected. He explained to those present why he declined the Gatling guns, saying the weapon would only impede their efforts if the troops had to chase down the Indians. There were some in attendance who felt sure the reason was Custer's unwillingness to share any of the glory of victory with another unit. Custer had often boasted that the Seventh Cavalry could single-handedly defeat the entire Sioux nation.

His next statement was a first and it stunned the group of officers. Custer asked for input from the men; something he had never done before. No one knew for sure if he was serious; they were so surprised that they remained silent. There was a touch of stuttering in General Custer's voice when he dismissed them (Custer tended to stammer when excited or under pressure). As they made their way back to their bedrolls the officers felt a strange uneasiness. Lt. George Wallace made a statement that has been quoted many times (I quote here from *A Good Year to Die*, by C. M. Robinson). He said, "I believe General Custer is going to be killed." When asked why by another officer he answered, "Because I never heard him talk that way before." While Heath lay dozing under the stars some of the officers made out their wills. Lieutenant Cooke was another such pessimist. He finished his makeshift will by remarking that he felt sure he would not survive the much-anticipated battle.

At three o'clock on the morning of June 23, the Seventh broke camp.

There was no reveille this morning as Heath was awakened. It was decided that the noise of the bugle might draw attention to them should there be any Indians in the area. Stumbling in the dark and with horses snorting, William mounted up with his company for another day's march. As they left camp this the second day out, a morbid sense of doom spread among the troops. Everyone felt it with Custer being unusually quiet by foregoing the usual banter between himself and his family members present. Along on this expedition were Custer's brother Tom, another brother Boston (who was serving as scout), his young nephew Autie Reed (sister Ann Reed's son who came along as a herder for a summer outing to fortify his health), and brother-in-law Lt. James Calhoun (husband of Margaret Emma Custer). If William Heath were the superstitious kind he must have felt very uneasy on this tense morning.

The column was broken down into smaller units and spread out so as not to kick up too much telltale dust. As they rode along the Rosebud River they began to come across signs of the presence of large numbers of Indians. An old campsite contained many lodges. Around burned-out campfires they saw piles of bones from meals eaten by lots of braves. The surrounding grass was eaten low by the grazing of herds of horses and their droppings could be seen everywhere. The trail they were following was six inches deep from ponies dragging lodge poles and was almost one-half mile wide. The increasingly nervous scouts took great pains to point all this out to General Custer, who insisted they keep moving.

At the end of the second day's march Heath had ridden a challenging thirty-three miles. The troops were almost six miles ahead of the supply train which was falling farther behind by the hour. It took them four hours to catch up when the column stopped for the night. The dust they endured while marching all day was more than just aggravating. It got in their eyes, burning them bloodshot. It sandblasted their ears. Despite bandannas over their faces the dust got in their mouths and made their throats raw. One can almost picture Billy Heath telling his fellow troopers how it reminded him of the mine dust when he toiled in the Black Hell. Not even the constant howl of the coyotes could keep Heath awake that night after such an arduous march. He had no way of knowing that much worse was awaiting them.

Up again before daylight, the Seventh Cavalry mounted up and pushed on. Custer knew the enemy wasn't far off and he was eager to catch them. The day of June 24 brought encouraging signs to Custer while his scouts became ever more worried. Broken branches along the trail were still green. There were piles of fresh horse droppings covered with flies everywhere. The Seventh was gaining ground. Eventually they came upon a large abandoned Indian camp. They found the drawing left behind by Sitting Bull. The scouts could hardly contain themselves as they considered this sign of powerful medicine and its consequences. In the sand was a drawing graphically depicting the end of the attacking troopers which the scouts correctly interpreted as their eventual demise.

This day's march had been even more brutal than the day before. Although they made twenty-eight miles this day (about five less than the day before) the terrain was especially rough. The men were suffering from saddle sores from the long rides. Heath's face, like the others, was swollen from sunburn as the day turned blistering hot. Deerflies attacked the horses, driving them at times into a frenzy and making them difficult to control. Buffalo gnats—tiny black flies—besieged the troopers, biting their eyelids and ears, causing them to swell up. The soldiers stank from both their own sweat and that of their mounts. When they stopped for the day the horses rolled in the dust to rid themselves of their tormentors. The men were afforded no such luxury and had to put up with their own insufferable smell.

Early that evening Custer met with his officers again. The officers were gripped by a growing sense of despair as their general clearly demonstrated a lack of confidence in himself. Some of the officers who did survive the battle later reported Custer to be unusually disoriented and in a depressed state during the entire march. They were also disconcerted knowing that they were overtly disobeying their orders to report back to command when they had information regarding the Indians' location. When they emerged from the meeting, something ominous happened. The standard banner was blown over by a sudden gust of wind. It was retrieved and replanted only to be blown over once again. Its effect on the men who witnessed it was one of an omen of certain defeat.

Omens, harbingers, and forebodings must have been running rampant throughout the Seventh at this time. William Heath's thoughts surely were now in contrast to those he had while camped outside the steamer *Far West.* Before marching in, his contemplations were likely tranquil ones. No doubt he chatted with his two fellow Schuylkill County friends as well as other members of the company. Excitement was growing with the anticipation of the coming chase. The men were confident as they looked around. There were thousands of them with adequate ammunition (including Gatling guns) and ample supplies of food and good mounts. Pleasant memories danced in Heath's head recalling the card playing, tale swapping, and the camaraderie developed by men up at reveille together practicing drilling and pulling guard duty as a unit. Now he could trade the hay pitching, wood chopping, and horse tending for chasing down the fleeing enemy and ridding the world of the red menace. No more escort duty, hunting expeditions, or shoveling horse manure. It was time to get to it.

Now, only hours before the battle, Heath's thoughts were probably anything but tranquil. The scene was vastly different from the pre-march one. Having endured what he had over the last several days and seeing the signs of imminent battle, we can almost surmise with certainty his mental state. Excitement was there but so was trepidation and uncertainty. They had all trained for this but were they ready? The sudden and uncharacteristic lack of confidence shown by their commander (which was being much talked about by the men) had to be unnerving. The growing nervousness of the scouts in the face of clear evidence of huge numbers of Indians close by began causing doubts. The exhaustion of the march, the intense heat, and the lack of food and rest were further eroding those previously secure feelings.

William Heath's singular responsibility to take care of everyone's mount, repair broken harnesses, and resecure thrown packs was a heavy weight for any one man. At least, with this preoccupation, Heath may have had less time than the others to feel that gnawing sense of fear and doom that began to fill the ranks of the Seventh Cavalry. God, how he would have liked to lie down that night for a good sleep. Alas, there would be no rest for the weary that last night together. Duty and their general were calling.

The troops were given a quick meal and the fires hastily put out. The men fell to the ground to catch their breath while awaiting further orders. Saddles were left on the tired horses and packs remained secured to the mules. Neither animals nor men were very happy. Farrier Heath was probably scurrying from problem to problem while the others rested. Horses were lame, mules were throwing their packs, and equipment had been broken or lost on the way. William Heath was most likely the busiest man in the company.

No sooner did they begin to catch their wind than Custer called for his officers. He told them scouts reported an Indian camp ahead with an estimated fifteen hundred braves. At 11 P.M. they were told to be ready to march in an hour. The general wanted to get as close as possible under the cover of night and conceal himself at the first light of day. Tripping and cursing in the dark, the soldiers mounted up. With mules braying loudly in protest, they pushed on through the pitch black. Troops had a hard time staying together in the darkness. They did the best they could by following the sounds of the horses' hooves on the rocks and the clink of equipment. The going was painstakingly slow; by daybreak they had made only seven or eight miles.

Custer had sent an already exhausted Lt. Charles A. Varnum ahead with some Crow scouts to a known high point called the Crow's Nest. At an elevation of one thousand feet he hoped to fix the enemy's position. Varnum reached the Nest by 2 A.M. and waited for daylight. At first light he strained as he scanned the horizon for sight of the rest of Custer and his men, but could see nothing. The Crow scouts told him to look for worms not horses (the huge Indian pony herd resembled a tangle of worms in the distance). Perhaps Varnum was just too weary. He had been riding almost nonstop going through one fresh mount after another for nearly seventy hours.

Scout Lonesome Charley Reynolds looked through field glasses and said it was the biggest pony herd any man had ever seen. Scout Mitch Bouyer was with them and agreed, noting it was usually like this when they all came together for the Sun Dance; then he added that the herd was far too big. The scout Bloody Knife was sent down from the lookout point in the Wolfe Mountains, where Varnum was located, to report to Custer. He told Custer there were too many Sioux and that it would take days to kill them all. His warning fell on deaf ears. Nothing could stop the general now. Custer marched the men the remaining ten miles to the Crow's Nest and climbed up to see for himself. By now the horses of the Seventh were nearly wasted. With tongues hanging out and eyes bulging there was more work for an already weary William Heath tending to their needs.

Bouyer and Reynolds pointed out to Custer where to look. The general remarked that he couldn't see anything that looked like herds of ponies. Bouyer angered him by replying if he didn't find more horses than he'd ever seen before in his life Custer could hang him. Lieutenant Varnum

Scout "Mitch" Bouyer. (Courtesy Denver Public Library, Western History Collection/Nast/X-31214)

advised Custer that he was certain the cavalry was spotted a short time earlier by a small band of Indians on horseback. Actually, a scouting party of Cheyennes and Lakotas had watched Custer's advance for some time the day before. They reported their findings to the Indian village. The chiefs did not think any white man foolish enough to attack such vast numbers as they had in their temporary village.

Mitch Bouyer made one last plea to his commander. He suggested that Custer get his outfit out of the country as fast as their exhausted horses would take them. The general didn't even acknowledge him with an answer. He climbed down from the Crow's Nest to his men waiting in a ravine below. Soon he was given more news that they had been discovered. A sergeant had gone back along the trail to get something that he had dropped. When doing so he came across several Indians working on a dropped box of hardtack and he was certain they had not missed seeing the trail the troops had left behind. There was no question now; the element of surprise was gone. This warning also went unheeded. Custer told his officers that he couldn't see anything from the Nest. He added that he doubted the others did either. What he did see was opportunity and he was determined to seize it.

His mind made up, Custer informed his officers that since they obviously had been discovered, it was essential that they attack at once instead of waiting until the light of day as he had originally planned. The condemned men of the Seventh Cavalry sat down for their last meal. No fires were permitted. A hastily thrown together fare of raw bacon, hardtack, and warm canteen water was all they got. While William Heath gulped down his food (if he had the time), he had to have begun questioning the wisdom of coming west. Could the death threats of the Mollies have been any more intimidating than this?

The Seventh Cavalry's destination, Little Bighorn, was still about thirteen miles southwest from its present position. The men were checked to make sure they had the one hundred rounds for their rifles and twenty-four for their pistols. With Heath in his usual position in Company L, the Seventh rode on as the sun rose high in the sky and the temperature climbed by the hour. Benteen saluted Custer and offered Company H to take the lead. Custer, stammering now, returned his salute and reluctantly gave him the advance position. They were on their way to legendary status in the pages of American history.

A short while into the march General Custer stopped the column and made a tactical decision that many historians say contributed greatly to his downfall. He divided his command into three separate battalions. He directed Reno to take Companies A, G, and M with about 112 men. Benteen was given D, H, and K totaling 125 soldiers. Custer took Company's C, E, F, I, and L with somewhere in the neighborhood of 225 troopers. Company B was given the task of guarding the supply train to the rear. This probably made them angry at drawing such mundane duty. In retrospect it saved their lives.

The pace quickened as they resumed moving. Surely the boy general was beside himself with anticipation of a victory. He envisioned this battle as the climax to a brilliant military career and perhaps even a springboard to the presidency itself. Both men and horses were well spent by now trying to keep up with an eager Custer. They crossed a divide, which brought them to a creek. Following the creek down a valley between hills on both sides they found themselves at the Little Bighorn River. At this point Custer ordered Benteen to take his men to some high bluffs several miles away. He was instructed to keep mounting the hills until he gained a vantage point from which he could pinpoint the location of the Indian village. Then he was to report back.

Benteen left at a 45-degree angle from Custer's position as instructed by his commander. He took no scouts with him and it was the first time in his life in this part of the country. Custer placed Reno and his three companies on the left bank of the river and, taking the right side with his contingent, ordered them all to proceed. The general came upon some unoccupied tepees and was checking them out when Indian interpreter Fred Girard (who had just reached the top of a nearby bluff) spotted some warriors riding away. He yelled to Custer words to the effect, "There are your Indians, General, running like devils." At this same time Reno noticed some twenty braves watching them from a distance safely out of range.

Without hesitation General Custer sent Adjutant Cooke to inform Reno that the Indian village was only about a mile away (actually it was more like four and a half miles away) and that he was to "charge the village and you will be supported by the entire outfit."* Historians differ on the exact time this took place. It is safe to say that it was approximately 3 P.M. when Reno and his men began their charge.

---

*From Stephen Ambrose, *Crazy Horse and Custer* (Garden City, N.J.: Doubleday and Co., 1975).

On the way Reno and his men hit Ash Creek at a full gallop. The mounts went crazy at the smell of water. They had not drunk in twenty-four hours and the late June heat was unbearable. They plunged in and began gorging themselves. Major Reno managed to restore order and forced them forward. By now he could see Indians ahead of him, fleeing. Some reports say at this point Reno sent back a messenger calling for support, saying the enemy was strong. If he did, it went unheeded. About seven hundred yards short of the village the horses could run no more and pulled up totally exhausted. Suddenly a force of about six hundred braves mounted a vicious attack on Reno and his men. Surprised and physically drained, Reno ordered his men to dismount and set up a skirmish line.

The minute he did this Reno and his men made the deadly transition from the hunters to the hunted. Momentum quickly swung to the Indians. Finding his position untenable the major ordered a retreat to a nearby stand of timber. Already he had lost men and horses. It is at this point that historians hotly debate whether or not Custer was aware of Reno's predicament. Most certainly he heard the shots fired. Some of Reno's men would later say they could see Custer and his men on a high bluff with the general waving his hat as if in approval (of what I don't know as he was already in trouble). Much more disturbing is a report aired on the History Channel recently regarding this controversy. In it a photographer by the name of Ed Curtis is supposed to have interviewed three scouts who were with Custer on that high bluff. They said they knew Reno was in mortal danger and asked the general to go help him. Custer is said to have replied no, let them fight, and led them away.

The initial onslaught of warriors mounted against Reno's position was soon joined by many more. They rushed Reno's horses, scattering some. They began working in from his rear, threatening to close all means of escape. Lt. George D. Wallace tried to convince one of the scouts to ride to Custer for help. He refused to go, saying it would result in certain death. Reno was losing men faster than acceptable. One soldier was shot through the stomach and fell from his saddle. Another came to his aid and, seeing that the gunshot was fatal, dragged him to some bushes, leaving him with a canteen to die. When he was found dead days later, his body had escaped mutilation by the Indians. I mention this only to illustrate that it was entirely within the realm of possibility for one to avoid discovery by the Indians in the confusion and disarray.

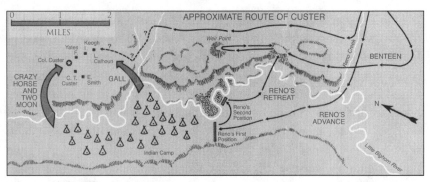

Exhibit 15. The Battle of Little Bighorn, June 25, 1876.

The situation was fast becoming desperate. As Reno sat astride his horse next to the scout Bloody Knife, he must have been wondering where the support was that General Custer had promised. A moment later a shot rang out, striking Bloody Knife squarely in his head, spreading blood and brain fragments all over Reno's face. Another trooper close by took a hit in the back of the neck. The bullet exited through his mouth, killing him before he hit the ground. Clearly they had to get out of there fast. Reno gave the order to mount up. They headed for the riverbank they had crossed earlier. The bank was a six-foot drop to the water and many horses balked at the intimidating height. The retreat backed up like a clogged drain and it was here that Reno's command took its worst casualties; Lonesome Charley Reynolds was one of them.

Somehow Reno and his remaining men made it across the Little Bighorn River to a saucerlike depression on a high hill. They dug in and, for the time being, were able to keep the Indians at bay from this lofty position. Before too long Benteen would arrive at Reno's position, asking him Custer's whereabouts. Reno replied that he didn't know. He had lost half his command and pleaded with Benteen for help. Benteen shared his men's ammunition with Reno's and the two forces held their position until dark. The sound of gunfire off in the distance told them another battle was raging somewhere to the north along the Little Bighorn.

Reno's attack to the south of the Indian village caught the Indians by surprise as they were occupied with watching Custer's larger group approach from the north. At first they retreated, buying time to evacuate women and children from the point of Reno's encounter seven hundred yards short of the village.

The village itself was in a peaceful slumber when the battle began with Reno's charge even though it was about three in the afternoon. The Indians had literally danced around fires all night until the sun came up. Many braves were asleep in their lodges after an exhausting night of dancing, singing, and pursuing the young women. When the word of the white man's attack was sounded they got the women and children out of the way and quickly assembled warriors and horses to meet the challenge.

The Indian village that Custer encountered lay spread out along the Little Bighorn River for almost five miles. The various tribes were arranged in their usual tribal circles (although a true circle of the entire camp was impossible due to the hugeness of it and the natural limitation imposed by the river) roughly according to the drawing in exhibit 15. The village included the Hunkpapa with Chiefs Gall and Sitting Bull, the Sans Arc with Spotted Eagle, the Blackfeet with Jumping Bear, the Miniconjous with Red Horse and Lame Deer, the Oglalas with Low Dog and Crazy Horse, plus the Yanktons, Santees, Brulés, Cheyenne, and Arapaho. As Bouyer and Reynolds had earlier told Custer, it was the largest gathering of Indians they would ever see.

After General Custer sent Reno on his way, he continued along on a path using his memory of a map drawn in the dust that morning by his scouts. They showed him the terrain all the way to the village, indicating where water would be. At first the five companies moved at a slow pace. When they came to a creek Custer's horse plunged forward to get his fill. Custer let Vic quench his thirst and then moved aside for the others to do the same. As William Heath and the rest moved in, the general cautioned them to not let the horses get waterlogged. The respite lasted a scant five minutes before moving on.

As the troops neared the Little Bighorn River, Custer picked up the pace and headed north along the east side of the river. They galloped now over a series of rolling ridges that looked like the exposed backbone of some giant animal. The commander sent out two scouts to the river's edge to check on the enemy. When they did not return he slowed down the pace and sent out the four remaining Crow scouts with instructions to do the same. Earlier that day Custer had an altercation with his scouts. At a rest stop he noticed them taking off their army clothes and dressing up in their traditional Indian garb. Custer wanted to know what they were doing and was informed through an interpreter that since they knew they were going to die that day, they would die as Indian braves and not as army soldiers. The irate general screamed at them to run home to their children and squaws.

When the four Crow scouts returned with confirmation of a large Indian village ahead, Custer picked up the pace again. He raced out in front of his men, urging his weary horse on. The men struggled to keep up. Some fell behind, as their horses just weren't up to the task. Custer's brother Boston's mount went down in an exhausted heap. Boston got him to his feet and drove him back to the rear to get a fresh mount from the supply train. He didn't want to miss all the action.

Custer had visual contact with the village now and sat on his horse for several minutes surveying the scene. He saw only women and children milling about, not knowing the warriors had raced off to meet Reno's charge. At this point he uttered the famous words that Heath would probably never forget, "Hurrah, boys, we've got them. We'll finish them up and then go home to our station."* He galloped for about three-fourths of a mile before getting close enough to the river to find there wasn't a good place to ford it. He pulled up and gave instructions to Sgt. Daniel A. Kanipe to ride back to the pack train and tell them to come quickly as there was a big Indian village ahead.

Putting his spurs to Vic, Custer charged across a rise and a draw to the right, his troopers as close to him as they could stay. He stopped and called Trumpeter John Martin forward. The Italian immigrant who had marched with Garibaldi was given both verbal and written instructions to go back and find Captain Benteen. He was to tell him to come quickly and to bring the ammunition packs with him. Martin rode off, going down in history as the last man to see Custer and the five companies of the Seventh Cavalry alive.

Martin's account of his ride is worth including here since it is an eyewitness report by the last man to leave Little Bighorn alive. This summary is taken from "The Story of Custer's Last Message" as told by Martin to Lt. Col. W. A. Graham and published in *U.S. Cavalry Journal.* Martin recalled that after being given the message, he was told to ride as fast as he could down the same trail they had just come. If he had time and was in no danger, he should return; otherwise, he was to stay with Benteen.

---

*From C. M. Robinson, *A Good Year to Die* (New York: Random House, 1995).

John Martin, trumpeter. (Courtesy Denver Public Library, Western History Collection/D. F. Barry/B-279)

Custer's Last Charge. (Courtesy Denver Public Library,
Western History Collection/Kelly/X-33659)

His horse was, in Martin's words, "pretty tired." Private Martin urged him on as fast as he could go. The last he saw upon leaving was the command descending into the ravine. The gray horses were in the center and they were in full gallop. On the way he ran into Boston Custer coming all out on a fresh mount. Boston pulled up and asked, "Where's the general?" Martin answered, pointing over his shoulder, "Right behind that next ridge you'll find him." At that the younger Custer dashed away to join his brother in death. A few minutes later Martin reached the top of a hill and heard gun reports. Glancing back he saw the Indians waving buffalo robes and firing on the Seventh. "They had been in ambush." Martin's account makes no reference either to seeing any men deserting the battlefield or encountering any such persons along the way back, the most likely avenue of escape for a deserter.

On the hill Martin reported he could see Reno's men, surrounded by a lot of Indians, fighting in the valley below and falling back. At this point he was spotted by some Indians who fired on him. "Several shots were fired at me—four or five, I think—but I was lucky and did not get hit. My horse was struck in the hip, though I did not know it until later."

The day was hot and he pushed his horse as hard as he could. He was not certain exactly where Benteen was or where to look for him, but he knew he had to try. Staying to the trail they had used, he eventually caught sight of Benteen. He tried to urge his horse on. The spent horse could go no farther. Benteen rode the last few hundred yards ahead of his command to meet Martin. Martin saluted and handed the message to Benteen, who read it and then asked where the general was. Martin replied that he had probably charged through the village by now. Then they saw the blood running from Martin's horse. Martin got a fresh mount and remained with Benteen until they came upon Reno. The major came out to meet them and exclaimed, "For God's sake, Benteen, halt your command and help me. I've lost half my men."

Martin finished his account by relating that at the time he'd been in America for only two years and this was his first Indian fight. He ended up serving in the U.S. Army for thirty years before retiring. After leaving the army Martin recalled, "My memory isn't as good as it used to be, but I can never forget the Battle of Little Bighorn and General Custer."

After sending Martin on his way, Custer divided his troops yet again.

Capt. Myles W. Keogh took command of Companies C, I, and L. Capt. George W. Yates had E and F. Historians differ on whether the general stayed with Yates or Keogh at this point. The two companies under Yates went as far as the river. Keogh and his three companies stayed on the higher ground of a ridge separating them from a drainage area to the north called Deep Coulee. Most students of the battle surmise that Custer stayed with Captain Yates along the river, for they rode about a mile or two within sight of Keogh looking for a way to find the north end of the village and sweep through it, scattering the Indians (most likely Custer's idea).

Yates and his troopers got to the Little Bighorn River at almost the center of the village. They were suddenly met with intense fire from a group of braves posted in the brush around them. After losing several men and seeing the Indians receiving more reinforcements who had been freed up once Reno took refuge on the hill, Yates was forced to withdraw from the river. The men had dismounted to form a skirmish line. Soldiers were holding their horses' reins in one hand while trying to fire with the other. The frightened horses were jerking the men around, making their return fire completely ineffective. Companies E and F sought sanctuary on a high ridge on the slope of Deep Coulee.

Long before Custer got to the north end of the village the Indians had seen him coming. The advancing column of blue troopers brought cries of warning from the Indians. Hundreds of braves now forded the river on their ponies and disappeared into the ravines on the other side.

Mitch Bouyer and several Crow scouts were some distance behind. They could see the ambush taking shape. Bouyer turned to the seventeen-year-old scout Curly and said, "You are very young, you don't know much about fighting. Go back, keep away from the Sioux, and go to those other soldiers, there at the Yellowstone. Tell them all of us here are killed."[*] Then Bouyer spurred his horse into a full gallop toward Custer and, looking back over his shoulder, yelled, "Save yourselves!" Curly and the other scouts heeded his advice and turned their ponies toward the hills in the direction of General Terry.

Indians by the hundreds hid in a ravine behind the hill Custer was marching across. The route of his march left him vulnerable on three of his four sides. The warriors struck with a savage vengeance, "like bees

---

*From Mari Sandoz, *The Battle of Little Big Horn* (Philadelphia: Lippincott, 1966).

Curly, Crow scout. (Courtesy Denver Public Library, Western History Collection/D. F. Barry/B-101)

swarming out of a hive" as one Indian account later recalled. While Custer, Yates, and the men of Companies E and F struggled to keep from being overrun, Keogh met with resistance from the Indians on the high ground between Medicine Tail and Deep Coulee.

At first William Heath's Company L and the other two companies returned fire from horseback. They took down several Indians and a number of ponies, causing the front riders of the braves to scurry for cover. Soon, they regrouped and began surging forward. Whooping at the top of their lungs and waving buffalo robes, the Indians spooked the soldiers' horses, making the rifle aims of their riders go astray. Then Keogh ordered one company to dismount and form a skirmish line, keeping the other two in reserve. Every fourth soldier grabbed the reins of three horses and led them to the rear of the action where they were secured.

Farrier Heath was probably one of these. Dodging arrows as well as gunshots (perhaps trying to get off some shots of his own), he did his best to lead away the terrorized horses. The action was frantic now, as Heath might have had to dispatch some badly wounded mounts while chasing down those that were running wild in the confusion. The threats of the Molly Maguires must have seemed small indeed to Heath at this moment compared to what he now faced. His thoughts may have turned to home and how he wished he were there. This very day was Midsummer Day back home in Girardville. Margaret and their two sons were probably enjoying the celebration along with all of the town's residents. There was good food to eat. The fireworks display widened his children's eyes, though not in the same way the fireworks around him widened his. Understandably, Heath thought he'd never see home again.

Another charge from the Indians dropped a number of horse handlers. Somehow William Heath escaped this volley. Horses stampeded, taking with them the ammunition in the saddlebags. The group of men on the skirmish line were now without mounts. Out of rifle ammo, they had to resort to their revolvers to return fire. Keogh ordered one of the reserve companies under Lt. James Calhoun to charge down the hill, temporarily scattering the Indians. The company dismounted and occupied a small ridge to hold their position.

Sensing Yates was in trouble and fearing the Indians were beginning to mass to his rear, cutting him off from joining up with Yates, Keogh moved his men north. They reorganized on a flat hill, now known as Calhoun Hill, overlooking the river. This is where Calhoun and most of Company L died fighting hordes of warriors. Heath, however, was not one of them. Calhoun Hill was part of a high ridge now called Battle Ridge that extended for half a mile. From here one had a grand view of the river below and the surrounding valley. From here is where the remnants of the five companies fought to the end.

If indeed General Custer was still alive at this point (as usual there is much speculation about exactly how, when, and where he died), he saw warriors pouring in from everywhere. He had to have known his situation was hopeless. One has to wonder if he thought about his boast that the Seventh Cavalry could defeat the entire Sioux nation as he found himself faced with the matchup he wanted.

By now the fighting between the soldiers and red men was growing more fierce by the minute. Keogh tried his best to hold onto Calhoun Hill despite the growing numbers of Indians pouring up the Coulee. Yates and whatever remained of his two companies continued fighting along the ridges. Wisely the Indians did not mount a frontal attack. Instead, they continued picking off the troopers from behind the cover of rocks and bushes, alternately charging and concealing themselves.

Heath's Company L was now in deep trouble. The men were besieged by Chief Gall coming in from the south, Chief Lame White Man from the west, and Chief Crazy Horse from the east. The men hastily formed a semicircle skirmish line on the forward slopes of Calhoun Hill. The horse handlers (which Heath, as farrier, would have been among) stayed with the mounts beyond the crest, no doubt trying to secure and calm the exhausted and terrified animals. Lieutenant Calhoun and L Company felt the fury of Gall and Lame White Man's warriors. Even the horse handlers began to draw heavy fire. A heroic charge by men of Company C (in an attempt to help the men of L Company) failed as they encountered intense fire from the red men and were driven back. Company L's circle of defense began to shrink as sheer numbers of the enemy overwhelmed them. Fighting became brutal as they went hand to hand with the enemy. This is where most historians believe the Indians suffered most of their casualties. Custer, most likely within view of the scene, could see his right wing go down in defeat. If William Heath had not thought of running before this he had to be seriously considering it now.

Crazy Horse delivered the final blow to Keogh. He forded the river and surrounded Keogh's position, squeezing the army of whites between himself and a band led by Chief Gall. The famous last stand took place at Custer Hill. Down to most of Company F plus part of E and a few sur-

Chief Gall. (Courtesy *Dictionary of American Portraits*, Dover Publications, Inc., 1967)

vivors from the other three companies, the men fought bravely. Some historians speculate there were a few suicides and some panic. Despite this, the Indian accounts testify to the valiant display of courage shown by the men of the Seventh.

A suicide squad of young braves administered the final coup de grâce. They charged the last pocket of men, sending the remaining horses scurrying as they engaged in hand-to-hand combat. The rest joined in as the end for the soldiers came in a tangle of them swinging rifles and Indians hacking. As the circle of soldiers grew smaller and smaller, the forty or so left were protected by a barricade of dead horses stacked up like so many sandbags. Just before the very end a handful of soldiers jumped over their wall of horses and ran down the slope for the river. Through the smoke and dust they charged in one last effort to save themselves from this nest of "swarming bees."

The Indians chased them down, hacking their living victims to pieces. Indian accounts say most if not all were caught and quickly dispatched. If anyone did get away, the Indians were sure the soldiers would perish for want of food, water, and exposure to the elements. Suddenly, after two hours of brutal battle (historians say it was all over by roughly 7 P.M.), a silence fell over the land. Custer and the men of the five companies lay dead—all except one. It was getting dark now and this lone survivor thanked God for night's protection.

The entire battle was a surprisingly brief couple of hours of brutal combat in which several hundred soldiers' lives came to an agonizing end. The scenes depicted in film and literature are graphic and ghastly. The initial clash between the troopers and Indians occurred on horseback when the Seventh's advance was boldly met by the warriors as they rallied from the village to protect their women and children. As the Indian numbers grew the Seventh was forced to set up the skirmish lines mentioned. With large numbers of braves arriving every minute the men had to stop kneeling to shoot and retreat to their horses. Firing from a stationary horse is tricky enough but doing it from the back of a terror-stricken animal is almost impossible. The horses' skittish movements caused the shots of the soldiers to go astray, depleting much-needed ammunition in the process.

As the cavalry set up its defensive positions each company's officer rode up and down the line giving orders and encouraging the men to stand firm. Not long into battle conditions severely deteriorated for the soldiers. Heavy smoke from thousand of rounds permeated the air. Dust from herds of horses and the feet of many men swirled around them. Visibility was reduced to zero. Blood-curdling cries from both men and animals shot in the most horrible of places pierced the air along the greasy grass. The smell of gunpowder was everywhere. Many troopers found themselves without mounts, leaving them unable to charge or escape. Such men banded together for protection, dragging their dead and wounded with them from position to position. When that no longer became feasible they left them behind. One can only try to imagine the disasterous effect this had on the morale of the men of the Seventh.

The soldiers were falling fast and furious at this point in the battle. Mark Kellog, the newspaper correspondent, and the mule he was riding were shot dead. As the conflict grew in intensity others began to fall. Death came to officers like Yates, Miles, Keogh, Calhoun, and the two surgeons with Custer. Members of the Custer family joined the general: Custer's two brothers, his nephew, and his brother-in-law. Civilians Mitch Bouyer and Lonesome Charley Reynolds died in hand-to-hand fighting with dozens of braves. The scout Bloody Knife died at the hands of his own kind. As the circle of white men shrank in size their resistance also diminished as if they were fighting in slow motion. The troopers' gun reports became audibly different as they went from using carbines to revolvers when the fighting became close. The final phase of hand to hand was brutal and brief. The Indians' war whoops filled the hills as hair-raising screams from the mortally wounded soldiers signaled the end.

In addition to their bows and arrows and warclubs the Indians were armed with a variety of white man's guns, many of which were superior to the troopers'. When the end came blood covered the earth and the stench of death was everywhere. Some experts subscribe to the famous last-stand scenario while others believe there were several last-stand efforts by the Seventh. Some have concluded the soldiers' defense was orderly, well thought out, and militarily sound. Others contend it was nothing more than mass panic.

Analysis of the battle site over the years includes studies of the location of bodies found, photography (both ground and air), archeological digs, and ballistic investigation. By looking at the shell casings under a microscope it can be determined which ones came from which guns. A study of these shot groupings lends some idea to the movement of individuals and groups of soldiers. The final conclusions are always hotly debated. It seems like the more information we gather the less certain we are what happened. Fitting it all into a neat little model eludes us. The only thing generally agreed upon is the devastating outcome. The controversy began in 1876 and has endured for 127 years.

The end for Custer himself is surrounded by the ever-present controversy that seemed to follow him throughout his life. Some believe he was killed outright early in the battle down by the river and his body hauled up to Custer Hill. Others say he was on the high ground to begin with and

Exhibit 16. Gen. George Armstrong Custer. (Courtesy Denver Public Library, Western History Collection/D. F. Barry/B-63)

ended up on Custer Hill for the last stand. Many Indian warriors claimed to have been the one to kill the general. Since he would have been extremely hard to single out in the melee, there is much doubt as to the truth of the claims. One Arapaho brave may have come closest to the truth when he said several Indians were responsible for his death. The Indian described Custer, dressed in buckskin coat, down on his hands and knees. He had been shot more than once and blood was coming from his mouth. It looked like he was watching the Indians all around him as a few men were sitting on the ground near him, too badly wounded to help him or themselves. Then, the Indian said, many braves closed around and he could see no more. A reasonable explanation as to why the Indians were somewhat confused as to the identity of the general was that he had cut his hair uncharacteristically short prior to this campaign. They were used to seeing his long hair to distinguish him from his men. Exhibit 16 is one of the last photos taken of Custer a short time prior to his death.

Before the dust had a chance to settle, there were acts of atrocities underway. Sitting Bull had prohibited such deeds, but his words went unheeded. Mostly women and youngsters moved among the corpses, looting and desecrating. The bodies were stripped naked. Jewelry, watches, anything of value to the Indians was taken. If a ring was too stubborn to come off a soldier's swollen finger, it was hacked off. Some women literally gutted the body of some soldiers who may have been responsible for making them widows. Hands and feet were cut off. Heads were bashed in. Hatchet marks scored the torsos.

Although naked when located, General Custer's body was not mutilated. He was found on his back with his hands folded across his chest as if napping. The only unnatural thing discovered was a sewing awl pushed into both his ears to admonish him for not listening to the chiefs who told him if he ever broke the peace, the spirit of the Indians would surely cause him to die. There were two wounds to his body, one near the heart and one in the left temple. Found dead with him were the four members of his family who accompanied him to the battle: brothers Tom and Boston, nephew Autie, and brother-in-law James Calhoun.

Author Peter Panzeri reports that Yellow Bird, the suspected illegitimate son of Custer, and his mother were present on the battlefield after the attack. As they surveyed the abattoir-like scene before them, she (with Yellow Bird standing meekly behind her) "gazed in silence at the naked pale-skinned body she had once loved."

❦   ❦   ❦

The system of communication in 1876 being nothing like it is today, official public word on the battle didn't appear until almost two weeks later. The *Far West* steamer carried the wounded remains of Reno and Benteen's men as hastily as it could to Fort Lincoln. Astonishingly, the boat traversed 710 miles down the Yellowstone and Missouri Rivers in a record fifty-four hours! On July 7 the closest major newspaper to the battle site, the *Bismarck Tribune*, ran the following multiheadline story:

MASSACRED
GEN. CUSTER AND 261 MEN THE VICTIMS.
NO OFFICER OR MAN OF THE 5 COMPANIES LEFT TO
TELL THE TALE.
3 DAYS OF DESPERATE FIGHTING
BY MAJ. RENO
AND THE REMAINDER OF THE SEVENTH.
SQUAWS MUTILATE AND ROB THE DEAD
VICTIMS CAPTURED ALIVE TORTURED IN MOST
FIENDISH MANNER

This is the very first printed account of the Custer massacre to appear to a shocked nation. It was dated July 6 but actually ran on the seventh. The article requires close scrutiny since its content (or lack thereof) has a huge impact on my claim that Heath survived. The headlines ask, "What will Congress do about it? Shall this be the beginning of the end?"

Bismark Tribune correspondent Mark Kellogg wrote, in what were to be his last words to the outside world, "We leave the Rosebud tomorrow, and by the time this reaches you we will have met and fought the red devils, with what results remains to be seen. I go with Custer and will be at the death." Prophetic words indeed as poor Kellogg met a violent end along with Custer and the five companies of the Seventh Cavalry.

Kellogg goes on to tell how Custer took up the march, hot on the trail of the Indians. His on-site report includes a reference to General Terry's urging Custer to take additional men (which the boy general declined) and some of the specifics of his orders. The account noted that Custer was to report by courier to Terry when locating the enemy so they could all get into position for the "final wiping out."

The rest of the *Tribune* article, written by someone other than Kellogg,

Unknown remains at the site of the Battle of Little Bighorn.
(Courtesy Denver Public Library, Western History Collection/Coffeen Schnitger/X-33173)

and taken from the accounts of those who were with Reno and survived, gives lurid details of what took place. It states that General Custer pushed on for seventy-eight miles with no sleep to reach the Indian camp, whereupon he attacked. Nothing more was heard from him until General Terry arrived to relieve Reno. Many a strong man is said to have wept at the glorious sight of Terry's appearance. It was then that Lt. James Bradley found Custer's dead body along with 190 members of the Seventh Cavalry. The report continues, "Of those brave men who followed Custer, all perished; no one lives to tell the story of the Battle.... Gen. Custer was shot through the head and body, seemed to have been among the last to fall.... Staff all dead—all stripped of clothing and many of them with bodies terribly mutilated. The burial of the dead was sad work but they were all decently interred. Many could not be recognized." I repeat, "many could not be recognized."

After the narrative section of the article came the headline "KILLED." The list that followed gave the name, rank, and company of all the fatalities. In all, along with a doctor, some civilians, and the scouts, some fourteen officers and 237 enlisted men were reported dead. Keep in mind that General Terry recorded finding and burying Custer and a total of 190 men. Do the math and it obviously doesn't match up. There is a discrepancy of about sixty men. If we include the civilians and scouts, it still doesn't add up.

Even more significant than the disparity in the body count is the singular fact that William Heath, Company L farrier, does not appear anywhere in the newspaper. He is not present in the list of those killed in action and he is nowhere else in the entire text. It is reasonable to interpret this important fact for what it is. William Heath was not among this initial list simply because his body was not found at the abattoir-like scene. It was not found because he was not there. Inexplicably, and beyond the army's belief, he had somehow managed to escape. His name was later added to the official Seventh Cavalry casualty list for the month of June 1876 as killed in action at the Battle of Little Bighorn in the erroneous assumption (held to this very moment) that no one survived that battle. Escape was simply unthinkable in the mind of the army, plus it added to the glorious mystique. To this day he is believed by the U.S. Army and countless Custer scholars to be buried in the common grave at the foot of the monument which has his name carved on it along with his fallen comrades. How this could be when the facts are in direct conflict eludes me, but I leave it to the army and the scholars to explain. If a man is declared dead and his body is present, there is no mystery or controversy as to his whereabouts. In the absence of a body and with ample proof that Heath lived afterward, I say there was in fact a survivor. Unquestionably, it is a significant error in the historical contention that no soldier escaped alive from Little Bighorn.

It is noteworthy here to point out that there are occasional references in the literature to Heath sometimes being listed as Kiefer. Indeed, there is a Kiefer listed as killed at the battle in the *Bismarck Tribune* account. Why this dual identification exists eludes me. Who originally made this assumption and on what grounds? William Heath never used the name Kiefer. The name Heath is the only one found on his enlistment records, recorded in the Seventh Cavalry report of casualties of the battle, and inscribed on the monument as can be seen in the exhibits. In fact, the Kiefer shown in the Bismarck account is listed as a private whereas Heath is always referred to as "farrier." Regardless, if one buys into this dual identification or not, the man by either name did not perish in the battle. There is indisputable evidence that he was alive and well for years after the struggle.

Exhibit 17. Little Bighorn Monument where the U.S. Army says William Heath lies buried. (Courtesy Ron Kensey)

The U.S. Army declared dead all the men of the five companies directly under Custer, including William Heath. Eventually the bodies of the fallen soldiers were interred in a common grave at the foot of a granite monument erected in 1881 (some officers were buried elsewhere at the request of family members like George Custer whose wish was to be buried at West Point). Exhibits 17 and 18 are photographs of the Custer Battlefield National Monument. One can clearly see the name of Heath engraved on it (eleventh from the top, left column). Exhibit 19 is a photocopy taken from the U.S. Army archives. It is the Seventh Cavalry's monthly report of casualties for June of 1876. It lists all men of the Seventh "killed in action with hostile Indians at the Little Bighorn River, Mt. June 25, 1876." Number 190 is none other than William Heath, farrier of Company L. Two spots down from him is the name of the colorful fellow Schuylkill County resident George Adams. These exhibits are conclusive proof that the U.S. Army believes Heath fought and died with General Custer at the Battle of Little Bighorn.

These exhibits are the most damaging to date for those who disbelieve that Heath was a survivor. Obviously these are conclusive pieces of evidence that the United States government, through the U.S. Army, claims William Heath among the killed in action on that June day in 1876. In order for them to do so, they had to be convinced he made muster that day

Exhibit 18. Close-up of monument. Heath's name is eleventh from the top left.
(Courtesy Ron Kensey)

and participated in the battle. Thus the army firmly establishes the necessary first part of a survival claim: that of the person in question initially being assumed killed in the battle.

At first the reports of Custer's demise were greeted with great skepticism by Generals Grant, Sherman, and Sheridan. They refused to take it seriously while they waited nervously for official word. Preposterous as it first sounded

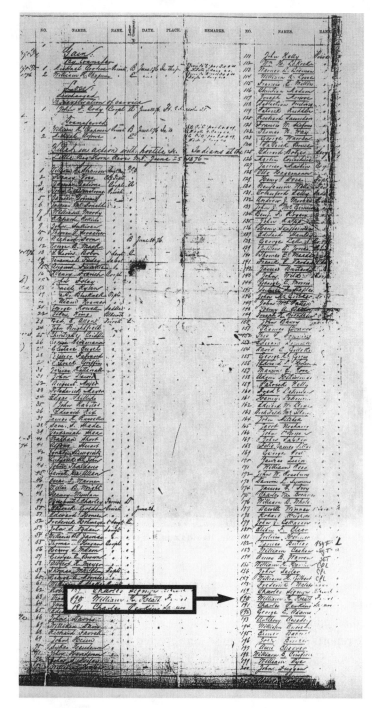

Exhibit 19. Monthly fatality report of the Seventh Cavalry for June 1876. (Courtesy U.S. Army Military Archives)

they soon learned of its truth. The following article appeared in William
Heath's local newspaper, the *Pottsville Evening Chronicle*, on July 7, 1876:

INDIAN MASSACRE
General Custer and his command slaughtered
FIGHTING LIKE TIGERS

Salt Lake, July 6—The special correspondent of the Helena, Montana,
Herald writes from Stillwater, Montana, July 2: Muggins Taylor, a scout for
General Gibbon, got here last night direct from Little Horn River. General
Custer found the Indian camp and 2000 lodges on the Little Horn, and
immediately attacked the camp. Custer took five companies and charged
the thickest part of the camp. Nothing is known of the operations of this
detachment only as they trace it by the dead. Major Reno commanded the
other seven companies and attacked the lower portion of the camp. The
Indians poured in a murderous fire from all directions, besides the greater
portion fought on horseback. Custer, his two brothers, a nephew, and a
brother-in-law were all killed and not one of his detachment escaped. Two
hundred and seven men were buried in one place and the killed are estimated at three hundred, with only thirty-one wounded. The Indians surrounded Reno's command and held them one day in the hills, cut off from
water, until Gibbon's command came in sight, when they broke camp in the
night and left. The Seventh Cavalry fought like tigers and were overcome
by mere brute force. The Indians' loss cannot be estimated, as they bore off
and cached most of their killed. The remnant of the Seventh Cavalry and
Gibbon's command are returning to the mouth of the Little Horn, where
a steamboat lies. The Indians got all the arms off the slain soldiers. There
were seventeen commissioned officers killed. The whole Custer family
died at the head of their column. The exact loss is not known, as both adju-
tants and the Sergeant Major were killed. The Indian camp was from three
to four miles long, and twenty miles up the Little Horn from its mouth.
The Indians actually pulled the men off their horses in some instances.
This is given as Taylor told it. The above is confirmed by other letters,
which say that Custer met fearful disaster.

On that same day the Schuylkill County *Miners Journal* confirmed
everyone's worse fears with this article.

GENERAL SHERIDAN'S OPINION—
THE STORY PROBABLY TRUE

Philadelphia, July 6, 1876—A reporter this morning had an interview with Lt. Gen. Phil Sheridan at the Continental, concerning the reported disaster. The general said: The only thing that makes me doubt the truth of the report is that too many are announced as slaughtered. I hardly see how it can be so, and hope it is not, although General Custer intended to march about as described.

After Sheridan gave this interview, a military dispatch arrived with an urgent message. The *Miners Journal* reported:

THE REPORTS CONFIRMED

Chicago, July 6—A dispatch confirming the report sent last night of General Custer's fight on the Little Horn River has been received at General Sheridan's headquarters.

Predictably, the press forged into the controversy surrounding the man who led the soldiers of the Seventh to their end. Disagreement over who was at fault was the order of the day. High-ranking officers accused Custer of disobeying orders and foolishly rushing into an ambush. Quotes like "foolish pride which so often results in the defeat of men" and "was a very selfish man, insanely ambitious of glory, had no regard for the soldiers under him, and attacked recklessly with men tired out from forced marches" appeared in print. Even Grant, who was no fan of Custer to begin with, was quoted as saying, "I regard Custer's massacre as a sacrifice of troops, brought on by Custer himself, that was wholly unnecessary—wholly unnecessary." Others came to his defense. One general said that he would have done the same thing, "The only way to fight Cavalry is with a dash—a charge. I don't blame him." Another blamed the mess on Reno for taking to the hills and abandoning Custer and his fellow soldiers. Reno demanded—and received—a Court of Inquiry in 1879, which decided in his favor. The debate continues to rage even now.

If William Heath's wife, Margaret, was aware of where her husband was stationed, she must have been devastated. After reading the *Miners Journal* and the *Pottsville Evening Chronicle* she had to be convinced that she was now a widow with two small mouths to feed. The description of the battle scene was graphic, dashing any hopes she may have had that her husband of only four years was alive. One can only imagine her thoughts as she read that the battlefield was "a sight to appall the stoutest heart." Yet, somewhere out there amid the rocky slopes and ridges hidden in a ravine and perhaps behind a small clump of bushes, William Heath was alive. He clung stubbornly to life, praying that somehow he would find a way out; that someday he might see his wife and sons again.

Chapter Five

CUSTER

MAN AND MYTH

I shall return to William Heath shortly. However, it is appropriate to pause here to take a closer look at General Custer and analyze his career and final actions.

After the Battle of Little Bighorn the telegraphs literally did not stop reporting the details for days. The public could not get enough about Custer; something that is still true today. He instantly became the personification of the brave mounted soldier leading the charge against the savage red men in the country's quest to extend its manifest destiny from sea to sea. His enduring image has always been one electrified with emotion. Whenever a bright rising star such as Custer falls from the sky, the word "legend" becomes overused. But, with the boy general it is almost impossible to abuse the word. He is truly an American legend. To many he is accorded godlike standing and justifiably so, they say. Like many before and since, his death guaranteed that which he sought while alive: permanent superstar status.

Since Custer's death there has been an endless, mind-boggling tribute to him. There are Custer scholars, Custer historians, Custer societies, military

strategists specializing in Custer books, motion pictures, theater productions, paintings, newspaper accounts, magazine articles, poems, songs, and
the like. Does the word "Custermania" come to mind? His image over the
years has basically remained untarnished and probably shines as brightly
now as it ever did. And, at the risk of blasphemy, I seem to be unable to
grasp the reason why. When one holds him up to the light of scrutiny, what
I see is not what most other people seem to see. I see a man with many of
the characteristics that society generally bemoans as undesirable.

At the risk of alienating a great many people, allow me to elaborate
my reasons. Custer's personal conduct, upon close investigation, reveals
character flaws too serious to overlook. First, there is the accusation of dishonesty regarding the granting of trading-post rights at military installa-
tions. There is some substance to the charges. Benteen (Custer's arch-
enemy in the Seventh) was not the only one with damaging details of kick-
backs and refusing to let soldiers frequent places not acquiescing to
Custer's demands. Lack of integrity also creeps in when one considers the
fraudulent stock scam Custer was involved in as well as a shady silver mine
deal out west where Custer sold his name and military influence for profit.

Next there is hard evidence that the general was guilty of infidelity
toward his wife, Libbie. A notorious womanizer, Custer was charged with
sleeping with the wife of one of his officers at Fort Leavenworth, Kansas.
His letters to his wife hint at being unfaithful. It is generally accepted that
he slept with an attractive Indian woman who was taken prisoner at the
Washita battle, and some believe he fathered the child she bore. His dishonesty is also obvious with his almost compulsive public lies. He lied about
how many Indians he killed in battle. He lied about his hunting exploits; he
lied to his superiors about his reasons for his actions on the battlefield, and
he lied to his wife about any number of things, suggesting their relationship
was not as loving a picture as many historians have painted. Finally, there
was West Point. Cadet Custer blatantly violated the West Point honor
system. Had he been caught, he would have been promptly ejected. It is difficult to view such actions as anything but deplorable.

General Custer's military conduct stands out in an even less favorable
light. The general openly criticized his superiors in the newspapers and
magazines of the day. He fed stories to reporters who hung around him
and sometimes he wrote the articles himself without using his name. Few

things will get you on the bad side of your boss like public criticism will. The great Indian fighter used his exploits (both real and imagined) in the Indian Wars to further his military career. He told people what they wanted to hear about the red savages and then made the most of the opportunities that opened up for him.

Nepotism is the only word to describe how Custer stacked the Seventh Cavalry with relatives: his brothers, his nephew, his brother-in-law. The result was that five family members were killed in one fell swoop. At least some historians concede that Custer's motivation in his last campaign may have had to do with unbridled political ambitions. Just back from suspension, his reputation suffering some damage, the general opted for a quick fix. If he could manage a fast dramatic victory over the Indians and get word back to the Democratic convention in St. Louis in time, he could have manipulated the delegates to nominate him for president. The testimony of an Arikara scout lends support to this theory. The scout said that Custer told him if he could decisively conquer the Indians in this battle, he could return to Washington where he would become the Great Father.

The mistakes at the Battle of Little Bighorn are numerous and damaging. Within hours after being reinstated by President Grant, Custer vowed to cross the very man who went to bat for him (General Terry). The history of the general leaving men behind on several occasions is well documented and needs no further comment. The same could also be said of his distasteful record of atrocities at Washita while in command. Although argued by Custer scholars, there is plenty of evidence to support a charge that Custer deliberately disobeyed orders at the Battle of Little Bighorn. First, he was not to have any newspaper men along on the expedition. Clearly he was in violation of this order. Second, General Terry fully intended for Custer to report back his findings and to avoid conflict until all forces were in position. Custer's failure to adhere to his superior's wishes definitely contributed to the loss of the men of the five companies as well as his own life.

Finally, Gen. George Armstrong Custer committed some inexcusable blunders immediately preceding the battle itself. He forced-marched his men for three days and a night in intolerable heat and then pitched exhausted men and horses into battle, greatly limiting his chance for success. He knew very little about the terrain on which he was to fight. This was a point that should have been corrected through reconnaisance by scouts. In addition, he totally ignored continued warnings from scouts and others with him (namely, Bouyer and Reynolds) that the number of Indians was much too large to engage with the force he commanded. You can add to this Custer's arbitrary dismissal of the news that he and his men had been discovered, thereby removing the element of surprise. Mix in the issue of splitting up his command not once but twice and it was a recipe for disaster. Last and particularly distressing to me, was the nonchalant way in which Custer ordered Reno to attack with the promise of full support that never arrived. Regardless of his feelings for the man, it is inconceivable that Custer could act in such an inhumane manner.

The question has been asked so many times. How could this have happened? How could a man like General Custer, who had an almost perfect record in the Civil War from Gettysburg to Appomattox, end up being annihilated at Little Bighorn? Most historians point to several theories, which I will briefly touch on. I will then offer one of my own.

Perhaps the most obvious and often overlooked reason for Custer's defeat at Little Bighorn is simply that on that day the Indians were the better fighters. Since the Indians seldom fought in an organized manner like the U.S. Army, that concept is hard to visualize. Nevertheless, the Native American victory at Little Bighorn can be viewed that way. For once the natives had superior numbers (an estimated six to seven times the number of soldiers). For once they chose to fight instead of fleeing like they usually did. In terms of fighting ability the Indians were brave and savvy warriors whose knowledge of undermining the enemy was not lacking. During this particular fight their motivation was supreme. They were led by the greatest chiefs of their time in a common cause they all shared, namely, the threat to their very existence as a race. By ignoring the obvious Custer paid the supreme price at Little Bighorn. The Indians were also in better physical shape, weren't tired, hadn't marched for days before, and they had fresh horses.

A second explanation frequently offered is that combat with the Indians was an entirely different type of warfare compared to the Civil War. In that conflict the tactics of the enemy were predictable. Both sides basically played by the rules reminiscent of the Revolutionary War brought over to this country from England. It wasn't so much a question of strategy as it was who had more of the better soldiers and more equipment. Custer's rise to fame in the Civil War was probably due more to that fact than to true military genius. Most of his stunning victories were the result of an all-out frontal attack (like his tactic against the Indian village). He was good at knowing when to hold them and when to fold them. He lost sight of this at Little Bighorn. Here the Indians changed from their usual practice of fighting for a while and then fleeing (to fight yet another day). This time they fought and stood their ground.

The Indians seldom mounted an all-out frontal attack. They hid behind cover, picking off their enemy. They hit and drew back to hit again and again. The terrain was different than any back east. It was rough to traverse with wagons in tow and rich with places to hide in ambush. The weather was vastly different from Gettysburg, Pennsylvania, or Virginia. The changes in elevation by the thousands of feet meant it could be intolerably hot one day and snowy or icy the next. Water could be plentiful at times or impossible to locate at others.

No one should have appreciated these differences more than the general, for he had been involved in the frontier Indian wars ever since the end of the Civil War. It is certain that Custer never truly got to know his enemy, the red man. He never immersed himself in their culture. He never acquired a regard for their way of life, how they thought, what they believed in. General Custer's view of the American Indian is a sad revelation of his way of thinking; a way shared by many others at the time. In one of his attempts at becoming an author, Custer wrote in his book *My Life on the Plains* that the Indians didn't live on the plains but infested them like a pack of coyotes. During a period of military inactivity, he lamented with tongue in cheek that while others were out there in the noble cause of killing Indians, he was just killing time. His opinion was that the Indian was not the noble red man depicted in literature but a savage in every sense of the word, with as cruel and ferocious a nature as any wild beast known.

Another popular explanation for Custer's foolish behavior leading up to the Little Bighorn disaster was his excessive desire to attain the highest office in the land. Blinded by political ambition, he would take any risk to attain his goal. Many historians propose that his thinking was clouded by the overpowering urge to become the youngest president ever. He needed political backing if he was to pull off a coup at the national convention that election year. He knew there was no better way to catch the delegates' eyes than to crush the Indians out west. Experience had taught him that America treated its victorious generals well. Dining with leaders of the Democratic Party in Washington and shaking hands with the big-money men in New York was fine. Still, Custer knew he had to have a great victory over the Indians to have a shot at his dream of the presidency. No price was too great to pay if it put him and Libbie in the White House.

All these theories make sense and are backed up by sound thinking on the part of historians and scholars. Any one or a combination of all three could be reasonable grounds for Custer's shocking defeat at Little Bighorn. Now I will jump into the fray with a deduction of my own. This is not presented impetuously. Thirty-one years of experience in the field of human psychology led me to suspect the etiology of Custer's actions. My investigation has revealed what may be construed as highly controversial results. Nonetheless there is hard evidence to back me up.

It is my educated opinion that George Armstrong Custer may have suffered from a borderline personality disorder. Granted, no one has interviewed him or administered any psychological tests. However, reviewing the anecdotal records offers enough evidence to support the possibility. Over the years the diagnosis and classification of personality disorders has undergone a continuous evolution. Criteria found in the *Diagnostic and Statistical Manual of Mental Disorders* is the basis for my theory.

A person with a borderline personality disorder shows instability in relationships, mood swings, and inflated self-image. His or her attitude toward other people varies considerably and unexplainably over time. The emotions of such an individual become erratic and can shift suddenly, especially toward anger. Borderline personalities tend to argue excessively and become irritable or sarcastic all of a sudden. They are impulsive, do things that are self-destructive, and have trouble establishing a sense of values or a choice of career. Such people can be difficult to diagnose because they frequently show signs of more than one disorder. Depression and mood swings are all part of the package.

More specifically, General Custer may have been a victim of a bipolar disorder more commonly recognized as manic depression. Keeping the general in mind, let us look at the seven critical symptoms of someone suffering a bipolar disorder. The first is an increased level of activity at work or sexually. Custer was a workaholic, often up before sunrise and continuing into the early morning hours. His ability to go day after day like this was as legendary as the man himself. Sexually, there are indications that he was not generally satisfied by one woman (recall the possibility of his infidelity referred to earlier). The second is unusual talkativeness, rapid speech, or stuttering. Any number of times during my review of the literature mention is made of Custer stammering when feeling tense or under some duress. He demonstrated this tendency three different times over the three-day period leading up to Little Bighorn alone.

Third is the racing of thoughts, causing one to make rash decisions, often in anger. Custer's anger was well known and well documented, especially among the men he commanded. His outbursts against his men were renowned; his disrespect for them apparent. Custer would tolerate no input from his officers. He accepted it if it came through proper channels, then promptly dismissed their ideas. When angry at his men he paraded them in shame in front of the others, had them beaten, threw them in jail, and even put some to death for desertion.

The fourth indication of manic depression is the need for far less than the usual amount of sleep. Custer routinely functioned on two or three hours of sleep daily, far below the normal requirement. His extreme bursts of energy can now be viewed in a different light—one of manic episodes in the absence of proper rest. Warning sign number five is a greatly inflated self-image, the belief that one has special powers, talents, and abilities. Over and over Custer confirmed his belief in this. His lifelong adherence to the "Custer Luck"—the belief that he could find his way out of any fix—is a good example. The nepotism he showed in stacking the Seventh Cavalry's officer ranks with family members shows his special feeling for himself. The court-martial for being away without orders, for using government property for personal use, and for forced-marching his men for his own selfish means of meeting Libbie at the train station is more than enough to establish that Custer felt above those around him.

Alarm sign number six is severe distractibility, rendering one unable to arrive at things like a set of values or a firm career choice. Custer's values were sorely lacking in every sense of the word. There was the stock scam, leaving men behind in battle, infidelity, the atrocities, and charges of dishonesty connected to the trading-post monopolies, to name a few. Custer was never really comfortable with his choice of a career. He is renowned for his military exploits, yet was constantly considering and even trying other occupations. He met with moguls in New York and thought about becoming a high roller on Wall Street. He sat with politicians in Washington and planned to switch to a political career. Never mind becoming a senator or representative, he was destined to become president. He tried his hand at writing, publishing his book *My Life on the Plains*, as well as numerous other articles (often written at 2 A.M. when everyone else was asleep). His dissatisfaction with his career choice is fairly obvious.

The seventh symptom is an excessive involvement in activities that will likely have very undesirable consequences, such as gambling, or decisions that could endanger one's life. The general was able to free himself from alcohol as he had promised Libbie he would. The gambling habit he was unable to kick although he kept vowing to his wife that he would stop. The decision to charge the Indian village at Little Bighorn literally flies in the face of reason, considering all the signs there were that warned to do so would result in dire consequences. Without hesitation Custer chose to go ahead regardless of the outcome. His disregard for the penalty he might have to pay smacks of a death wish. Custer suicidal? Unthinkable! But perhaps not. The combination of the seventh symptom coupled with the depression mentioned in the account of the march in has been known to lead to suicidal thoughts. Sometimes these thoughts are openly expressed while at other times the person entertains them, saying nothing to those around him. There has been speculation before on this likelihood by Custer scholars. They point out the unusual situation of Custer's body being found with his favorite hunting rifle and a spent shell casing from it at his side. Nowhere is the word "suicide" used, but the insinuation is clear. The motto of the Seventh Cavalry—"Save the last bullet for yourself"— takes on new meaning.

Interspersed among the aforementioned seven symptoms would normally be occasional bouts of depression. Without elaborating, we have seen several incidents where the general was, by the accounts of all those present, suffering from depression. I cannot prove that George A. Custer was bipolar; neither can my critics say with certainty that he was not. The American Psychological Association will state that if a person shows signs of any four of the seven symptoms it is justification for a diagnosis of bipolar disorder. Custer showed definite possibilities of legitimate symptoms in seven out of seven (eight if you are counting the depression). Today he would be treated by state-of-the-art methods like psychotherapy, electroconvulsive therapy (yes, it is still used and is still found to be very effective), and drug therapy.

If I am correct in my theory of Custer's malady, then an apology to him is in order. I have been very critical in my judgment of both the man and the myth. If he was in truth suffering from a mental disorder, I have been unfair. If the evidence is credible enough to support my contention, he needs to be looked at in an entirely different light: that of being mentally disabled and, at times, not completely responsible for his actions.

The sole credible investigation into the psyche of Custer that I have been able to locate is *Custer and the Little Bighorn: A Psychological Inquiry*, by Charles K. Hofling, M.D. The author of this work arrives at the conclusion that George Custer suffered from a narcissistic personality disorder. This "mirror personality" theory is offered as an explanation as to why Custer constantly felt the need to surround himself with people (such as Libbie, family, and close friends) who would repeatedly reinforce his much-needed desire for encouragement and unrestrained admiration to bolster his self-esteem.

Hofling also points to a disturbing pattern of relationships with both men and women. With older men who possessed rank and authority Custer was boyishly compliant. Anyone of the same age or younger and at or below his rank Custer treated with disregard, doing as he pleased. Women with less status than he were regarded as sex objects to do with as he wished. Hofling alludes to a report that Custer suffered from venereal disease supposedly acquired from his casual relationships with some New York showgirls.

Dr. Hofling further states that birth order and its circumstances helped to contribute to Custer's disorder. Custer's parents both brought children from previous marriages into their union. For three years he was their only child together. Eventually, younger brother Nevin came along and being sickly he constantly needed everyone's attention. This trauma affected Custer's self-esteem, resulting in behavior like earning demerits at West Point, gambling, and the flamboyant striving to be someone greater than he was. When shamed (such as was the case when he was suspended by the president right before the battle) he reacted with a surge of activity designed to seek glory and to erase the negative image that had attached itself to him.

In conclusion Hofling states that when General Terry (a man of rank and power in Custer's psyche) dropped the bombshell that he was going to ride with Gibbon and not Custer into battle, it threw Custer into a state of psychological limbo. Unable to decide whether to become compliant or rebellious, he became confused and depressed, losing the ability to make sound decisions in the forthcoming battle. Even taking into account the "permissible inferences" that Dr. Hofling relies on to arrive at his analysis, I consider his a far greater reach than my premise of bipolar disorder. It would seem that the experts will need to sort this one out.

One of the most interesting and insightful looks into the life of General Custer is Robert Utley's *Cavalier in Buckskin*. It is a good biographical source filled with eye-opening anecdotes. At the conclusion of his book Utley offers his assessment of Custer's actions at Little Bighorn. In spite of my hypothesis regarding the possibility of a personality disorder, I disagree with many of Utley's appraisals of the general.

Utley states that the troopers of the Seventh Cavalry were exhausted "no more than usual." I seriously doubt that those soldiers were used to three consecutive days of hard march climaxed by an all-night march the day before the battle. The excessive heat and its effect on the men must also be taken into account. Robert Utley also contends that Custer did not disobey Terry's orders. I disagree. Perhaps, if nothing else, the spirit and intent of those orders were violated by Custer. Utley further states that the general was wholly justified in following the Indian trail over the Rosebud divide instead of continuing up the Rosebud as Terry advised. I would argue the contrary, especially once Custer saw the signs and was forewarned by his scouts that the Indian numbers were so great.

Utley avers that despite all factors, Custer's decision to attack on June 25 was sound. It was not. He knew next to nothing about the terrain, was sadly outnumbered and knew it, showed little understanding of the mental state of his enemy by never considering the possibility that they might fight instead of flee, and he divided his command prior to attacking. Custer failed to make a single sound move prior to the battle. Robert Utley also makes the statement that Reno failed Custer. I would contend that it was quite the opposite.

Lastly, Utley is very critical of Reno's retreat to the hill thereby freeing up Indians to attack Custer's position. This defies logic. Why would anyone expect Reno to risk his life and the lives of his men to relieve Custer when Custer promised him full support and failed to deliver? Keep in mind that Reno had no idea where Custer was (certainly not right behind him as pledged) and he had precious little time (being in the midst of a heated conflict) to consider opting whether or not to go to Custer's aid.

Robert Utley writes that renunciation of the glorious image of Custer is both unthinkable and impossible. Why not? Aside from his bravery (which I do not question), Custer charged himself and his troops into mass destruction in a display of arrogance and selfishness that cost the lives of several hundred men. I fail to see the merits of the high esteem in which Custer is held. Saying his demise was due more to bad luck than bad judgment is nothing short of romantic rationalization.

Chapter Six

AND THEN THERE WAS ONE

I f, as I contend, William Heath did survive the Battle of Little Bighorn, how did he do it? How did he manage to avoid the carnage and escape detection, allowing him to flee on that fateful day (for the evidence presented will show that the question is not *if* he got away but only *how* he accomplished the feat)? That is a question that will probably never be satisfactorily answered; least of all in the minds of critics. However, at the risk of falling victim to the charge of supposition (of which I myself was so faultfinding at the outset of this book), I will attempt to surmise the possibilities. Again, I reiterate that none of my theories negates the fact that he did survive, but only posit how he may have achieved it.

I should begin by first commenting that in doing my research I have come across numerous references to stories by men claiming to be survivors of Custer's Last Stand. I have been unable to find anything of substance actually published. One man, E. A. Brininstool, is reported to have collected some seventy accounts by men claiming to have survived. To date all such claims have been discredited as concocted stories made up by men for their own self-aggrandizement. Even the most open-minded

Custer scholars seem to be unwilling to consider the possibility that one, or possibly more than one, might have escaped. William Heath never sought notoriety. He never spoke publicly about his ordeal and only sparingly to his family. I can only hope that we have not become so enamored with the legend and mystique of Little Bighorn that we refuse to accept a credible and verifiable account like Heath's. Questioning something we have held onto for a century and a quarter can be hard, but would the discovery of a legitimate survivor somehow negate the courage and valor of the men who died there? I don't think so. Without debating the morality of their mission, the men of the Seventh Cavalry died with honor. The fact that Heath lived through it does not detract from it one iota.

Although all white accounts disdain any possibility of a survivor, the Indian stories present an interesting and tantalizing contrast. *Lakota Moon* (1977), by Greg Michno, is a captivating study of Indian interviews that leaves the door open to someone living through the ordeal. In this book Cheyenne warrior Wooden Leg recalled that during a charge on the soldiers a mounted soldier suddenly appeared from the ridge behind them. The man was galloping his horse as fast as he could, trying to run right through them. Wooden Leg figured he was in hiding until the Indians passed by. A band of warriors chased him but Wooden Leg lost sight of them as they rounded a curve at the top of a hill. "He guessed that they must have caught and killed him." "Guess" is the key word here.

In another recollection of the battle Little Hawk, an Oglala chief who fought aside He Dog and Crazy Horse, remembered a man on horseback trying to escape. He was almost out of range when a lucky Indian shot knocked him off his horse. It was assumed the man was killed, but what if he wasn't and that man was William Heath? Indians Brave Wolfe, Rain in Face, Yellow Nose, Flying Hawk, Fears Nothing, Two Eagles, and Lone Bear all saw soldiers fleeing to the river and felt that none of them made it alive. However, the Indian Waterman said a few did get to the river, surmising that they were dispatched by the braves patrolling the area.

Custer in '76: Walter Camp's Notes on the Custer Fight (1976), edited by Kenneth Hammer, also gives the reader food for thought. An unnamed Arikara Indian who was serving as an enlisted soldier for the army reported a strange occurrence. He had been following Custer's charge to the village, lagging behind on an exhausted mount, when he came across two white soldiers on foot whose horses had collapsed. He saw five Sioux warriors surround the men and later concluded that they were both killed. Is it so inconceivable that one of them may have gottem away?

Camp, a Custer historian who interviewed survivors of the battle, relates an interview with the Indian Foolish Elk who fought at the Little Bighorn. Camp asked Elk if any whites were taken alive into the village and tortured. A somewhat annoyed Foolish Elk wanted to know why such a question was asked of him. Camp replied that the bodies of some eighteen men could not be found. Elk answered that they would have found them if they looked hard enough. Here were eighteen men unaccounted for. Could not one of them have been a survivor? In another Camp revelation, the Indian named Flying By said that after the battle he found two soldiers on the Rosebud nearly starved to death. He also said he heard of another man eating frogs to stay alive. Flying By does not say what became of these men.

Finally, Walter Camp retells an interview with Lt. Charles Roe from Colonel Gibbon's Montana Column who said following the battle, while camped on the north side of the Yellowstone River opposite the mouth of the Rosebud, he discovered the bodies of a cavalry soldier and his horse. He believed them to be the remains of a man who escaped from the Custer fight. The site where Roe found the remains is between eighty to eighty-five miles from the battle. If someone could make it that far, why is it so unbelievable that someone could have made it all the way out?

Let us begin with the most obvious possibility and the one sure to be proposed by detractors. Desertion is a possibility that can't be discounted but there is also no evidence to support it. Personally I do not believe Billy Heath deserted prior to the battle. I say this because, aside from the five companies of men who died, there were many men under both Reno and Benteen who lived to tell their stories. Anyone going over the hill prior to the splitting of the command would have been seen and reported. Men would have been dispatched after such a runner to bring him back or shoot him. I have not come across a single account of any such occurrence.

After Custer split his command with Reno and Benteen the march in to the Little Bighorn was swift, leaving little time to think about taking to the hills in strange country. By now every man in the Seventh Cavalry knew there was a huge number of Indians all around him. To a man they must all have been certain that their best chance for living through the impending ordeal was to stick together, several hundred strong, and follow the lead of their famous and experienced commander. To bolt at this point was to sentence yourself to near-certain death. By sticking together at least there was a chance.

Finally, there is Heath's history of being a company man. In spite of all the technology, statistical analysis, and sophisticated tools of today, the single best predictor of an individual's future behavior is how he performed in the past. Heath served with the Seventh through the winter of 1875–76 under brutal conditions. Men were deserting left and right during that time. If Heath were the deserting kind he would have been long gone when spring came prior to Little Bighorn, and this book would never have been written. Before his army days Heath showed his loyalty to the company back home in Schuylkill County as well. His pattern for sticking it out had been established.

One could easily argue, "But he ran away from the Molly Maguires." Although this is true, I believe he did so under entirely different circumstances. The Mollies knew exactly where to find him and left a clear message what they intended to do to him. By staying put he risked almost sure death. By staying put with the Seventh Cavalry he stood the best chance of living. Another very important factor involving the Mollies that certainly weighed heavily on Heath's mind and no doubt influenced his decision to leave was the harm that could come to his family. The Mollies took no prisoners. If they tried to shoot Heath or burn his house down and his wife and children got in the way, his family would pay the price. Obviously William Heath was not willing to risk this, and it is clear to me that, under what I consider a different set of circumstances, the charge that he had a history of running would be unfair.

A second chance occurrence is that once the fight began, when the skirmish line was set up, farrier Heath was busy taking three horses at a time to the rear to secure them. As the battle raged and it appeared they were about to be overrun, Heath may have decided to dash for safety. Mounted on horseback he could have attempted to charge right through the line of braves closing in from the rear while the worse of it was off to his left. He may have even been in the company of Sgt. James Butler, who may have been trying to do the same thing (or the dash could have been to go for help). As you may recall, Sergeant Butler was found totally isolated from the rest of the dead (see exhibit 15 on page 120 showing the drawing of the battlefield), touching off questions that have never been answered as to how he ended up where he was found. Is it possible the two men were together and Butler fell victim to the Indians and William Heath miraculously made it through?

A third scenario worth discussing is this one: William Heath fought on, staying to the bitter end. Somehow he made it to the final cut of the last forty or so men left alive. From behind their breastwork of dead horses and with precious little ammunition left, these men struggled to stay alive. With fallen comrades all around them and the Indians surging at them from all sides, they had to know their fate was sealed. In one last superhuman effort to cheat death, a handful of them charged out from behind their makeshift fort of horse carcasses and made for the river below. William Heath was one of them. The Indian accounts verify that this happened. They go on to say that most if not all of the fleeing bluecoats were run down and killed. They said if anyone did get away he would surely perish from the lack of means to provide for himself plus the ravages of the elements. Apparently they were wrong.

Once he reached the river with braves in hot pursuit Heath would have attempted to conceal himself. The braves knew he was there somewhere and kept searching. The escapee may have hid behind the cottonwood trees. When an Indian got too close he may have slipped quietly into the Little Bighorn River and submerged himself. He sought refuge wherever he could find it. Bushes, tall grass, fallen logs. Alternating between land and water he would have distanced himself from his would-be killers. By now it was almost 8 P.M. Thankfully the sun had disappeared behind the high hills and darkness was creeping over the land. One can almost picture two or three braves closing in on Heath's position about to discover him when an outburst of yelps from the hill distracts them. The celebration gets under way in full force. The last soldier has been slain and the victory festivities begin in earnest. Not wanting to miss out on the ceremonial dancing and chanting, the braves depart leaving the scared, wounded soldier to what they figure is certain death anyway.

Night could not come soon enough for William Heath. His relief at the total blackness was tempered by his lack in ability to travel in the dark through rough and unknown terrain. It would be foolish to risk death by falling over an unseen cliff after making it this far. Still, the urge to move may have prompted Heath to stay in the river, working his way along its southeastward course. Exhausted, soaked to the bone, and freezing cold now that the sun had set, Heath's movements were painfully slow. He stopped only when his body demanded it and then only briefly. In the distance there were the muted sounds of the Indian's victory dance to remind him to keep moving.

As soon as there was enough light Billy Heath would have instinctively moved, trying to put as much distance as he could between himself and the Indians. He need not have worried, for they were busy harassing Reno's position and then getting ready to pull up stakes to avoid the army's retaliation. At first he moved rapidly, feeling a sense of urgency. Eventually, he would have slowed his pace over time as the lack of food and water took its toll.

It is not known how many days he traveled or what distance he covered. My guess is that he headed in a southwesterly direction. If he had gone north he would have run into General Terry marching from the Yellowstone River and been found. A westward movement would have taken him to the Indians; due south might have brought him to Fort Kearny and General Crook. We know that none of these occurred so a southwesterly trek makes sense. This route is in the direction of the Bozeman Trail and was reachable in several days by foot.

The ground he must have hiked over was unforgiving. Beyond the low-lying river were miles of grassy hills sometimes covered with buffalo. Heath probably had no gun, and if he did, he may have had no ammo with which to kill anything to sustain himself. He would have kept an eye on the buffalo. They might run away or stampede directly at him. No doubt he gave them a wide berth. Sloshing through the water-filled gullies on these plains he would have encountered swarms of mosquitoes, flourishing in the stagnant water, and gnats and ticks living in the high grass that covered the plains. They have been known to drive an animal wild with frenzy, running until it literally dies from exhaustion while trying to escape their incessant biting.

Closer to the mountains the land becomes drier and covered with trees. Climbing to a high point to get a bearing Heath would have seen an eternity of treetops. With no clue where he was or where he was going, he had to be disheartened. Yet, he moved on. The first night out must have been even worse than the night he hid in terror from the Indians. He had started this journey after three days' hard marching plus the all-night march. He had nothing to eat, may have been wounded, and was out in the blistering sun. By nightfall he was desperate for rest and found little. Freezing from the sudden drop in temperature, soaking wet from fording streams, and with hunger gnawing at his belly, true sleep must have been impossible. The night air was filled with the howls of wolves and the yips of coyotes. What he wouldn't have given for a fire to warm himself and a piece of hardtack to gulp down.

Daylight brought more of the same. Heath most likely tried to stay close to watered areas. When finding them he had to drink deeply, filling his belly since he had no canteen and knew not when he might drink again. Sometimes he would be relieved to find water only to discover that it was too alkaline or brackish to drink. At other times, driven by thirst, he gulped down water that minutes later threw him into the throes of diarrhea and retching. How he could have endured such conditions is hard to fathom. Continuing on his way, Heath's pace was getting slower all the time as he crawled across sand ridges, stumbled down ravines, and climbed over deadfalls only to come face to face with more of the same. Walking in a straight line was impossible. Only the birds had the luxury of traveling that way over this terrain.

Along the way, if he was lucky, he may have eaten some edible berries or wild plums to fend off starvation. Venturing closer to the mountains brought the fear of bears, for this was grizzly country. The Indians feared and worshiped this great bear, the largest carnivore on the continent. The foothills of the mountains had their own special obstacles. The rocky slopes were slippery in spots, sending a climber tumbling backward after getting halfway up. The slopes were covered with rocks and boulders of all sizes and shapes. Many of them were hiding places for the western rattler, a deadly enemy if stepped on. Overhead Heath could see turkey vultures circling. He feared they would alert the Indians to his presence or, worse yet, a savvy old bear who knew what their circling announced. The wayfarer Heath kept pushing himself on. Forcing his way through scrub brush as he went, the bushes tore at his heavy wool uniform, ripping it to shreds and lacerating his skin. The smell of blood drew more mosquitoes. His misery had to have been unbearable.

Every time he crossed a stream the weary traveler had to stop to pull off the leeches that amazingly got under his clothes to start sucking him dry. Every time he stumbled on a boulder, threatening to break a leg, he gritted his teeth. After several days of this William Heath had to be dangerously close to dying: blistered feet from walking untold miles in wet boots, second-degree sunburn from exposure, starving, dehydrated, perhaps wounded and bleeding, delirious, and growing weaker by the moment. Would anyone blame him for giving up? Maybe, in fact, he did. Conceivably at some point, he lay down and waited for death to rid him of his misery. At least he could have sought consolation in the fact that he wouldn't suffer the mutilation the others had at the hands of the enemy.

Were such the case I would not be telling his story. Incredible as it sounds, William Heath was rescued. Oral family history relates that a passing family of settlers in a wagon discovered him and hauled him aboard. The exact location and date that they picked him up is not known. They could have been on their way out west to settle. It's even possible they were returning East after failing to find what they were looking for. Many a settler came home after finding the trek west too inhospitable and finding reprisals from the ever-present Indians too much to bear.

The family that picked him up from death's door may have looked much like the one in exhibit 20. Whole families in Conestoga wagons were crossing the country at this time. The wagons had canvas-covered tops and were drawn by as many as four oxen or horses. These prairie schooners could haul up to three tons of belongings and usually were filled to the hilt. Every conceivable item of need could be found onboard the wagon: food, water, furniture, clothes, seed, plow, shovels, rakes, and axes to name a few. A milk cow and an extra horse or two might be tied to the back. The settlers considered it a good day if they made twenty miles. The roads, when they could find any, were in horrible shape. The jolting of the wagon over rough ground could make you black and blue. The creaking of the wooden cart could put you to sleep until the next lurch threatened to break an axle, something that could prove fatal to a family out in the middle of nowhere. The riders had to put up with the gagging smell of sweat from the oxen as well as the constant blanket of dust that followed them everywhere.

They suffered with intense heat in the day and shivered through the cold night air. Herds of buffalo were all around them, drawing the flies that attacked the settlers. These hardy people endured sandstorms, driving rains

Exhibit 20. A Mormon family rests in front of two Conestoga wagons.
(Courtesy Denver Public Library, Western History Collection/X-11929)

and hail, and snow in the higher elevations, which could occur any month of the year. At night they looked for a likely campsite, one with water and grazing for the stock. They ate their meals over an open fire, then quickly put it out and prayed the Indians had not seen the fire or the smoke.

The surname of the family that plucked Heath from this godforsaken land was Ennis. The woman's name was Lavina. For years the descendants of the Heath family, especially Deborah Brumbaugh, have searched for offspring from the Lavina Ennis family to corroborate their story. To date they have been unsuccessful. Perhaps this book will reach the eyes of some member of that family who will recall the story of one of their forebearers saving the life of a man called William Heath in 1876.

More than likely when Lavina saw the condition of Billy Heath she figured he wasn't going to make it. However, being a stout-hearted woman of good Christian faith, she tried her best. Knowing the limited medical knowledge of the time as well as the unavailability of medicine, it is impressive that she did try and just as impressive that Heath responded. Using the crude methods available to her, she nursed him through the winter of 1876–77. The winter is on record as being long and brutal. The Ennis family spent it with the extra mouth of Heath to feed somewhere out in the uncharted West. Specifics on the extent of his injuries as well as how much Heath told the Ennis family about how he came to be there will probably forever remain a mystery.

Details of Heath's stay with his rescuers are nonexistent. We can only surmise what took place on that wagon in the middle of nowhere. It is not clear if the Ennis family was alone on their journey or in the company of other wagons. Heath did not speak of it often and he did not provide his family with many specifics. Perhaps it was too painful a time to recall. Being helpless as a child and totally dependent on the kindness of strangers can bring mixed feelings. Reluctance to talk about it is understandable. One also has to consider that William Heath may have felt guilty about not finishing his term of enlistment with the army. Whatever the reason, it is a puzzle that will forever remain unexplained.

The winter out west that year following the Little Bighorn battle is on record as being brutally cold. One can picture the billowing canvas top of the wagon buffeted by the cruel winter wind. Although it could be closed off at both ends in bad weather with a drawstring, it was still icy cold inside. The canvas, treated with linseed oil to make it waterproof, may have blocked the wind but did nothing to keep the inside of the wagon warm. Despite being loaded to the hilt with everything from furniture and food to farm tools, the Ennises found a spot in the wagon to keep the injured soldier. For a time he might have drifted in and out of consciousness. Lavina must have been a patient woman for, despite the odds, William Heath lived.

Lavina's husband deserves some credit here also. He had to go off in search of fresh game in order to feed his family and this extra mouth. When doing so he put himself at great risk. If several miles from the wagon he were to fall prey to the Indians or get bit by a rattler or thrown from his horse they would all perish. A fire had to be kept going at all times. A supply of wood had to be gathered for it. If no wood could be found he may have had to collect buffalo chips to serve the purpose. When a winter storm hit he probably had to construct a lean-to with a roaring fire in front for there was no way to exist inside the wagon under such conditions. Heath had to be moved each time they did this to keep him alive. The Ennis man surely tried to keep guard at night against animals and savages. He was a busy man finding water, caring for the livestock, crossing dangerous rain-swollen rivers, and doing his best to avoid dysentery and cholera, which would only add another burden to his wife and jeopardize them all.

The mystery of where the Ennis family was going will always intrigue me. My theory is that they were traveling along or near to either the Oregon Trail or the Bozeman Trail when encountering Heath. The Oregon Trail passes some two hundred miles to the south of the battle site and it is not inconceivable that Heath made it that far. The trail ran from Independence, Missouri, to Oregon in a zigzag northwest direction. It came through the prairies of Kansas to the Platte River in Nebraska. Along the way it went close to the Black Hills of the Dakota Territory on to Fort Laramie, Wyoming and then across the Rockies at South Pass. The trail undulated like a snake, moving northwest then southwest and then northwest again and so on. It was a well-established route west. By 1850 some forty-four thousand pioneers made the journey west. Seven years earlier Marcus Whitman took the first large-scale wagon train (about one thousand people) all the way to Oregon. A more likely possibility for finding Heath was along the Bozeman, for it was less than fifty miles from Little Bighorn, easily within reach either on foot or on horseback. Even though the Bozeman was not in use by 1876 it is possible that an occasional family chose it rather than the more popular Oregon Trail, hoping to avoid attacks by Indians (especially if their destination was Montana).

If you had the sense of adventure and the intestinal fortitude to head west on one of the trails in 1876, you had to leave Missouri no later than late April or early May. You had to at least scale the Rockies no later than mid-October. If you failed to keep this timetable you died. The Rocky Mountains cared not who your family was back east or how many of you there were or even how badly you wanted to complete your journey. Getting caught in the mountains when winter arrived meant death. Basically you had four good months to log two thousand miles. You could not allow yourself to be slowed down by trivial things like marauding Indians, indescribable bad weather, mud up to the wagon axles, swollen rivers, broken bones, stampedes, and the like. William Heath's recuperation was certainly anything but a bed of roses under such conditions. While he healed, the soldiers out west resumed the Indian campaign. Sitting on their horses and singing "Forty Miles a Day on Beans and Hay," they endured minus thirty degrees and heavy snows in their pursuit of the enemy.

While William Heath mended, the U.S. Army, bolstered by the full support of Congress, took its revenge for the crushing defeat it suffered at

Little Bighorn. The army's retaliation was something the Indian chiefs both feared and predicted after the Battle of Little Bighorn. On July 17, 1876, the army began its search-and-destroy mission. On that date Col. Wesley Merritt caught up with about one thousand Cheyenne warriors who were on their way to join up with Chief Crazy Horse. At War Bonnet Creek near Fort Laramie in southeast Wyoming the two sides struggled until the Indians were the first to withdraw. They returned to the reservation as the white man wished and fought no more. On September 9, Gen. George Crook attacked a large Sioux village in the vicinity of Slim Buttes, in the northwest corner of South Dakota. After Chief American Horse was mortally wounded he agreed to surrender if his braves were spared. He died watching his men leave for the reservation. Slowly and surely the U.S. Army was closing the dragnet around the Indians. This time there was no escaping the white man's retaliation.

By late October Col. Nelson Miles and a contingent of the army caught up with the famous Chief Sitting Bull. Miles attacked the chief and his followers at Cedar Creek, Montana, which is northeast of Little Bighorn along the Yellowstone River. Following two days of fighting Miles scored a victory with the surrender of two thousand braves. Sitting Bull was not among them. He escaped and headed north, fleeing into Canada; his power to lead his warriors in battle against the white man was forever diminished.

In early January 1877 the same Colonel Miles relentlessly tracked down Chief Crazy Horse. Finally, on January 8, Miles engaged the enemy. The two-day battle took place on the snow-covered slopes of Wolf Mountain, in southern Montana near the Wyoming border. In the midst of a fierce winter storm the Indians took a brutal pounding by the artillery Miles wisely brought along. The Indians were the first to withdraw, scoring a victory for the army. When spring came, some three thousand Sioux reluctantly surrendered to the white man at the Red Cloud Agency Reservation in northwest Nebraska. By September Crazy Horse was dead. He was killed in a scuffle with his army captors on the reservation.

Over the succeeding years the government of the United States embarked on a master plan to deal with the Indian problem that reflects sadly on its character. The strategy was to turn the Indians into farmers. Their language, customs, and religion were to be eliminated through repressive

Buckskin Lodge; Sioux prisoners; General Custer's battle flag, captured at the Battle of Slim Buttes; and the officers in command of the charge. (Courtesy Denver Public Library, Western History Collection/S. J. Morrow/X-31747)

education on the reservations. They were moved from one place to another whenever it suited their white captors. They suffered from droughts, starvation (when the government failed to adequately supply them as promised), and epidemics from the white man's diseases (e.g., smallpox and alcoholism, to name only two). The once-proud race of Native Americans was decimated, and it is a wonder that they were able to survive the American holocaust. Thanks to lethal doses of bullets, booze, and bacteria, the government's master plan had forced the Native Americans into total submission. By 1890 the Indians ceased to be a military threat and had merely become a paternalistic problem to be handled at the government's whim.

Sometimes it takes 125 years for history to repeat itself. Such is the case of the presidential election of 1876. The campaign for president that year bears a startling resemblance to the one we recently witnessed in the George W. Bush versus Al Gore tussle of 2000. As William Heath lay languishing somewhere in the middle of no-man's-land after Little Bighorn, the conventions to pick our next president were under way. The comparison of it to the Gore/Bush battle is uncanny.

By the end of Grant's final year in office (1876) the last in a series of scandals struck. Secretary of War William Worth Belknap was found to have pocketed twenty-four thousand dollars from selling the rights to supply the Indians with their annual provisions. The House of Representatives voted unanimously to impeach him. Belknap escaped prosecution, saving face when Grant accepted his resignation with regret. Despite all the furor, Grant's supporters urged him to seek a third term. It seems the parasites were reluctant to lose their generous host. The House put an end to the era of malfeasance by reinforcing the two-term resolution ending Grant's try for four more years.

The most popular man in the Republican stable at the time of the convention was the magnetic statesman James G. Blaine from Maine. Possessing a remarkable memory for names and faces and an excellent speaking voice, the congressman mesmerized his audiences. He appeared to be a sure bet for the nomination in spite of the Democrats' (who were keenly aware he had never served in the Civil War) retort that he was invisible in war and invisible in peace. Shortly before the convention, disaster struck Blaine. Charges were made that Blaine had improperly secured a land grant for one of the railroads and was financially rewarded for his efforts.

When the Republican convention met in Cincinnati, Blaine's flame was still burning bright enough for a nomination from the floor. Before a full vote could be taken, a rumor circulated that there was something wrong with the gaslights and it was too dangerous to turn them on. The meeting was adjourned for the day and when reconvened the next day, cooler heads prevailed. A compromise candidate was chosen. Blaine's brief shot at the presidency slipped away from him with the nomination of Rutherford B. Hayes.

Hayes, the bearded unknown, was the perfect compromise candidate. He was far from brilliant but smart enough. He had fought in the Civil War where he had been wounded several times. Although he only attained the rank of major general, he would bring in the veteran vote. He was in favor of cleaning up the corruption in government and restoring sound monetary policy to the economy. Furthermore, in his favor was the fact that he hailed as a three-time governor from the electoral vote–rich state of Ohio. He was chosen to carry the fight to the Democrats.

The Democrats, meanwhile, met in St. Louis to find their champion. It was here that General Custer had planned to bring news of a stunning victory in the Indian campaign, hopefully stampeding the delegates into nominating him for the presidency. Sadly for Custer, the plan took a nasty detour in Little Bighorn country. Politics being what it is, the Democrats nominated (with little regret for Custer) the well-known bachelor governor of New York, Samuel J. Tilden. The intellectual Tilden was rather common in appearance. He was frail, sickly, and weak-voiced, the latter giving rise to deriding comments from the GOP like "whispering Sammy." With all the gusto of a presidential campaign, the race was on.

The Democrats cried it was time to put an end to the scandalous Republican rule and reform the government. Republicans announced proudly to the public the clean past record of their man. "Hurrah for Hayes and honest ways" was their slogan. The Democrats cried, "Turn the rascals out," trying to make Tilden's name synonymous with reform. The contest was as heated as it was enthusiastic.

In shades of the 2000 election the returns poured in to an eager nation. The Democrat Tilden seemingly had the election won. He quickly amassed 184 of the then necessary 185 electoral votes. With some twenty electoral votes in doubt in several states due to some voting irregularities, the Democrats felt sure he would come up with at least the one needed to push him over the top (especially since he polled about 260,000 more popular votes than his opponent).

The Republicans, in the midst of preparing their concession speech, spied opportunity and rallied. They quickly sent a delegation to the states in question to "oversee" the tallying of the disputed votes. Not surprisingly, the Democrats did likewise. The end result was two sets of returns from the states, with one set favoring the Republicans and one set the Democrats. Charges and countercharges flew of forgery, intimidation, and impropriety on both sides. Weeks went by with no end to the deadlock in sight. Investigation into the language of the Constitution brought no solution. The document mandates that the electoral votes be sent to Congress and in the presence of both bodies they are to be opened by the president of the Senate. What was not specified was who was to count them. If counted by the head of the Senate (a Republican), the outcome was obvious; if by the head of the House (a Democrat), equally obvious.

A constitutional crisis loomed. Some kind of compromise was the only way to avoid it. Some Democrats literally began military drills in anticipation of forcefully seating their candidate. Congress thought it had the solution when it passed the Electoral Count Act in early 1877. The act called for a counting commission of fifteen members from the Senate, House, and Supreme Court. There were seven Democrats and seven Republicans with the fifteenth to be Justice David Davis of the Supreme Court. He was an Independent but had definite Democratic leanings. Just when the Democrats felt they had the situation in hand, Davis resigned from the bench to take a Senate seat. Much to the Democrat's chagrin, all the remaining justices were Republicans. They were back to square one again.

In February the electoral votes were opened before a fully attentive House and Senate. In due time the votes from the state of Florida (one of the states with disputed votes) were brought forth with its two sets of returns. The commission voted eight to seven in favor of accepting the Republican version of the votes: the Democrats could see where this was going and vowed to filibuster until hell froze over if necessary. At last, behind closed doors, an agreement was worked out. The Democrats agreed to allow Hayes to take office in return for concessions in cabinet positions, taking federal troops out of several Southern Reconstruction states, various internal improvements, and the like. Historians generally acknowledge that both sides performed numerous polling atrocities and shared the blame for the crisis. It is often referred to as the election that the Democrats stole only to have the Republicans steal it back from them. Despite taking office under the cloud of illegitimacy (Hayes was dubbed for a while "his fraudulency"), President Hayes's term was one of integrity and met with some degree of success in cleaning up the government.

Chapter Seven

HOMEWARD BOUND

W hen winter broke out west in the spring of 1877, William Heath had recuperated enough to travel. He must have departed with mixed feelings from the family who saved his life. Without doubt he was happy to be healthy and on his way. It must have been a bittersweet good-bye. Before leaving, oral family history as reported in a 2001 *Leighton Times* newspaper article entitled "Custer's Last Stand," reveals William promised his faithful nursemaid Lavina that if he ever fathered a daughter, he would be proud to name her after the caring woman.

I have made numerous references to oral family history giving us details, though sparsely at times, about William Heath's life. Naturally some could question who made these statements and whether we can believe them. Oral history is not documented and therefore usually open to doubt. Thoughts like was the story embellished, did it even happen in the first place, or isn't it strange that such details are only coming out now 127 years after the fact come to mind. Oral history has been collected and used in retelling history for thousands of years. Before the written word it was the only history we had. Much of our ancient recorded history, although now written down, originated as oral. Someone else witnessed an important event or overheard something and related it to someone who wrote it down, making it recorded history. Using their own words to do so they may have unconsciously colored the event in the process. The power of the printed word may be overstated and oral history greatly underestimated.

To an amazing extent the oral family history regarding Heath has generally been corroborated by the documents presented in this work. It does not appear to have been enhanced, but merely handed down in earnest over several generations without any motive of fame and fortune. When I report that family history tells us William returned home after the battle, the tax records, census records, and the like bear this out. In doing the research I found no glaring examples of the oral family history being grossly out of line with the documented facts.

So, who were these storytellers? They were the sons and daughters of William himself. It was they who passed the knowledge of what William and Margaret told them on to their children who in turned passed it on to theirs. The one most often quoted by the present descendants of Heath (his great-grandchildren) is Lavina. Fittingly, Lavina is the daughter of Heath who was named after the woman who kept William and this story alive. Unfortunately Lavina wasn't privy to much detail since William seldom spoke of his past experiences. Possible reasons for this behavior are offered in chapter 8.

However, Lavina did say that William told the family he served out west with the Seventh Cavalry and was immensely proud of that service. It is also Lavina who related to her children the story of William finding the death threat on his door. These and other tantalizing details have been passed down to family members and only recently have been revealed to anyone outside the family. The details appear to be both genuine and unenhanced. It might be waxing to the dramatic to paint a scene with Lavina sitting around the fireplace on a cold winter night telling her children of the perils of their grandad out west. More likely she took the opportunity at family gatherings to keep the memory of William alive by retelling what she knew about her dad. It is highly unlikely she had any concept at all as to the historical significance of her words.

As Heath prepared to depart the Ennises, his last belongings were being auctioned off by the army since he had been declared dead. As noted in

Men with Custer, by Kenneth Hammer, "The final statement of William H. Heath, signed by Lt. Winfield Edgerly on March 3, 1877, indicated his personal effects were sold at auction for $5.00." At the time, the army records indicated Heath was owed $1.14 for tobacco and $15.00 for clothing.

We can only imagine what thoughts Heath entertained as he prepared to travel. He might have considered staying out west to rejoin his unit. He could have thought of seeking adventure and fortune in the gold fields. He would have been safe choosing either of these for no one (least of all the army or his family) was looking for a dead man. Evidently the option to return home must have been the stronger as that was his destination.

No one could have gone through the ordeal William Heath had just endured and remained unchanged by it. For the rest of his life he would never be quite the same after his experience at Little Bighorn. One can only speculate about what motivated him to return home instead of rejoining his unit. Heath may have been moved by an intense longing to see his wife and children again after emerging from the jaws of death. It is conceivable that he was experiencing guilt over being the only one to survive from five companies of soldiers. The psychological anguish of being the lone survivor of a battle can be mentally devastating. Heath could have been persuaded by the idea that he'd done his patriotic duty, put his life on the line, and now it was time to go home.

William Heath most likely contemplated all of these thoughts while suffering the turmoil of posttraumatic stress disorder, a legitimate likelihood under his extraordinary circumstances. The malady has many unpleasant side effects. Spontaneous reexperiencing of the nightmarish event induces great stress. Someone with this problem will experience a kind of numbness of emotional responses to others around them. Such people are anxious, angry, and subject to depression. The ability to concentrate and make sound decisions is seriously impaired.

This psychological disintegration of soldiers was first recognized as early as the Civil War. At that time, the army termed it "nostalgia," meaning a severe melancholia brought on by an extended absence from home and family. By World War I it was more popularly known as being "shell-shocked." Today, post-traumatic stress disorder (PTSD) is accepted as a real form of mental disorder resulting from enduring an extreme life-threatening experience. Human beings need to establish some sense of worth and meaning in the way they live their lives. When one questions the intrinsic meaning of a horrifying past experience, failure to confront it can bring on anxiety and depression. Such may very well have been the mental state of Heath when making his decision to return to Schuylkill County. Home was where his wife and children were. Home was where he hoped to sort out his thoughts and come to terms with what had happened to him.

As this lone survivor of Custer's Last Stand headed for home, had the army known of his existence, it would have no choice but to label him a deserter. His enlistment papers indicated that he signed up for a five-year term of service. By not returning to his unit when he was able, he subjected himself to the charge of being absent without leave (AWOL). Heath had barely put in about seven months' time until the Battle of Little Bighorn. Before being overly critical of his AWOL status we should keep in mind what he went through and what his mental state was when deciding not to return to the army. Unlike many who simply went over the hill at the first opportunity, Heath's choice came after the struggle and probably under great mental strain.

It is interesting to speculate on what mode of transportation he used to get back to Schuylkill County. Unless the Ennis family provided him with a horse, he would have been left with no other means than to walk. Maybe he was fortunate enough to hitch a ride with passing wagons or a stagecoach for part of the journey. Another plausible prospect is that he could have hopped a train when he found the chance. When the train slowed for a curve or to climb a steep mountain grade, Heath might have jumped aboard a freight train and concealed himself behind the cargo. Or he could have worked for a few days to get enough cash to buy a ticket for at least part of the journey. Details of these interesting activities went to the grave with him.

The fact that in the spring of 1877 he did return to Girardville is easily verified by the Heath family history, as we shall see. I'm sure you can imagine the scene when the knock came at the door of the Heath household in Girardville. Picture the look on Margaret's face when she opened the door and to her astonishment, like Lazarus returned from the dead, she saw her beloved husband, William, standing before her. No doubt she flew into his arms, beside herself with excitement. Two very young children, John and William, must have been bewildered. William Jr. in particular was born one month after William Sr. enlisted in the army and never even saw his father before this. The child, not yet a year old, must have been frightened by the combination of his mother's reaction and this strange man's arrival at his house.

Billy Heath was back in his home again. As Margaret fussed over him the only telltale sign of a visible scar denoting he had endured some brutally hard times was the absence of the top half of his left ear. I'm sure she quickly decided she could live with that. Heath was reported to be very self-conscious about the missing part of his ear and always wore a scarf around his neck in an attempt to hide it. How he explained it to those outside the family we can only guess. Surely the questions flowed from Margaret's mouth. How? Where? What? and a dozen more in rapid succession. The house must have rocked with the celebration of his return. Late into the night the sounds of laughter and excitement filled the rooms. That evening, after so many miles and so many life-threatening nights, William went to sleep in his own bed with his own family around him.

If William Heath needed time to readjust, he didn't take much of it. He had a family to care for and he had been gone too long. William returned to doing the only thing he knew. He went back to working in the Black Hole as a miner. Exhibit 21 shows Heath back on the Girardville tax records for 1877. Perhaps working in the mines didn't seem so bad now. It certainly beat what he faced out west. Through it he could make a decent wage and provide for his family. The worst of the violence in the coalfields had subsided and the Mollies had enough troubles of their own without coming after him to settle an old grudge, especially now that he was no longer working for the hated Coal and Iron Police. Life here would never be as exciting as out west. No doubt that was just fine with William Heath.

This exhibit is the first of many records affirming the second necessary part of a survival claim—that of proving the dead man is indeed alive. William was back working in his former occupation of miner and paying his taxes like everyone else. This confirms what oral family history retold all along—that Heath returned alive from the Battle of Little Bighorn. Clearly, if a man is dead he can't be going to work every day in his hometown and paying his taxes. Dead men don't pay taxes.

There was some excitement in the air around the time William returned home to Schuylkill County. On May 4 the trial of the six men

SCHUYLKILL COUNTY, ss.

M. W. Fehr, Louis Blass and Patrick Collins

Esquires, Commissioners of the County of Schuylkill, in the Commonwealth of Pennsylvania, To *A. B. Schlessman*

Assessor of the *Borough of Girardville*

in the County and State aforesaid, ... GREETING:

YOU are hereby required to take an account of all Freemen, and the personal property in your *Borough* made taxable by law; together with a just valuation of the same, and also a valuation of all trades and occupations made taxable by law; and of this you are hereby required to make a just return within thirty days from the date hereof, noting in the same all alterations in your *Borough* occasioned by transfer or division of real property; and also noting all persons who have removed since the last assessment, and all single Freemen who have arrived at the age of twenty-one years since the last triennial assessment, and all others who have since that time come to inhabit in your *Borough* together with the taxable property such persons may possess, and the valuation thereof, agreeably to the provisions of the law. You are hereby further commanded. diligently to inquire after and take an account of all real estate which since the first of May last may have passed from persons dying seized thereof, otherwise than to or for the use of father, mother, husband, wife, children and lineal descendants, born in lawful wedlock, and to set out the same in schedule attached hereto; and also of any estate or estates which may have come to your knowledge from any executor, administrator or otherwise, together with a fair and just valuation on the same, according to the market price thereof, and make return thereof to us with the list of taxable property. You are hereby further commanded and directed to re-assess, between the periods of the triennial assessments' all real estate which may have been improved by the erection of buildings or other improvements, subsequent to the last preceding triennial assessment—subject to appeals, as now provided by law. You are also required to furnish a list to the County Commissioners of all male persons in your district between the ages of 21 and 45 years, subject to militia duty, with the number of taxables of any independent School district or districts in your *Borough*

THE DAY OF APPEAL will be held at the *house of Louis Blass* on the *2nd* day of *May* next.

IN WITNESS WHEREOF, the said Commissioners have hereunto set their hands and affixed the seal of the said office, at Pottsville, this *26th* day of *March* A. D. 187*7*.

M. W. Fehr

Louis Blass

Patrick Collins

       Commissioners.

Attest: *E H chartet*   Clerk.

YOU, *A. B. Schlasman* elected Assessor for the *Boro of Girardville* in the County of Schuylkill, do swear, that you will diligently, faithfully and impartially perform the several duties enjoined on you as Assessor of the said *Boro* by the above precept, without malice, hatred, favor, fear or affection, to the best of your judgment and abilities.

Sworn and subscribed before me, this *2nd* day of *May* A. D. 187*7*.

A B Schlessman

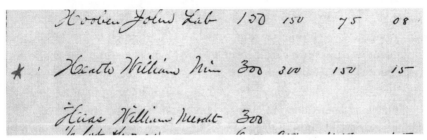

Exhibit 21 (*above and facing page [Tax Cover Page]*). **Girardville Tax Records 1877.**
(Courtesy Schuylkill County Courthouse Archives)

reported to be members of the Molly Maguires who were accused of murdering policeman Benjamin Yost began in nearby Pottsville. The trial held the attention of everyone in the county. The major newspapers in the eastern part of the country followed the trial, reporting daily on its progress. It fast became apparent that the outcome would depend on the testimony of one man: James McParlan, the undercover agent who infiltrated the Mollies, hired by Frank Gowen.

McParlan was born in Ireland and came to the United States sometime in the 1860s. He began life in the United States by toiling at hard labor in various jobs. Soon, his keen intellect was recognized by officials of the Pinkertons who immediately hired him for a special task they had just been given. He was employed by the Pinkerton Agency for twelve dollars per week to conduct industrial espionage on the Mollies. McParlan was precisely what Mr. Gowen had in mind. He was as Irish as Paddy's pig with red hair atop his short stocky frame. McParlan had a beautiful tenor voice, which he used to sing the many Irish ballads of the day, including "Molly Maguire of Donegal." He was as quick with his fists as he was with his wits and, to top it off, he was Catholic.

Despite fitting the stereotype of a Molly, McParlan apparently showed no sympathy for their cause. He was hired to do a job and he swiftly set about it. He was even sent back to Ireland for a short stint to study the workings of the secret society where it had its origin. Even Frank Gowen did not know his true identity, only his initials.

By October of 1873 McParlan showed up in Schuylkill County disguised as a tramp. Pretending to be intoxicated, he stumbled into a local bar in Pottsville and did an Irish jig while singing a tune. He immediately won over everyone in attendance and was invited into the back room for a game of cards. In a short time he discovered one of the men cheating, resulting in a barroom-clearing brawl outside. James easily bested his opponent and instantly became one of the boys. The secret agent told the men he was on the run from a murder in Buffalo, New York. That was just the thing they wanted to hear. He had them in his pocket by now.

One year later, taking the oath on his knees for the fee of three dollars, James was initiated into the Ancient Order of Hibernians in Shenandoah (a town right next to Girardville). As a trusted member of the inner circle he could now gather the specific information for which he'd been hired. The agent witnessed firsthand what had been suspected by Gowen, that the Molly Maguires existed mainly to protect its members and seek revenge if any of them were threatened in any way. People who crossed them were marked for execution. Others had their heads bashed in by clubs and their tongues torn out. For lesser infractions—only speaking ill of them—the Molly Maguires had the person's ears cut off. The fear of the Mollies was as real as the deeds they performed on anyone who dared go against them.

Gowen's patience paid off in September of 1875. Back in the city of Philadelphia James McParlan told everything to his employers. He gave them over three hundred names of Molly Maguire members along with their addresses, occupations, their place in the hierarchy of the organization, and their illegal activities. It was the best money Frank Gowen ever spent.

James McParlan was walking a thin line between life and death. Eventually it was a Catholic priest from Pottsville who grew suspicious enough to turn detective. He became convinced that McParlan was a plant and, when the word reached head Molly Black Jack Kehoe of Girardville, he was marked for execution. So good was McParlan at convincing people he was the real thing that it probably saved his life. The night the hit was to take place a rival of Kehoe's, certain that McParlan could not possibly be guilty of being a snitch, whisked him away in the dead of night just as the assassins were readying their weapons.

McParlan dropped out of sight and remained hidden until it was time to testify against the six men on trial for the policeman's murder. Up to the task as usual, he did a credible job of testifying for the prosecution. When the defense attacked him he held his own against charges that he played a major part in inciting the Mollies to violence and murder. There was some credence to the accusations, but the outcome of the trial was pretty much predetermined since it had been orchestrated by Frank Gowen himself. All six defendants were found guilty and were hanged from the gallows in Pottsville at the end of June, almost one year after the Little Bighorn battle. The newspaper reported the men were awakened at 4:30 A.M. More than likely they were already wide-awake. They had coffee and waited for their wives and family to arrive at 6. Time was allowed for tearful last good-byes. All six, being Catholic, were taken to church for mass and communion promptly at 7. They were led from church directly to the gallows. Their legs were bound with leather straps and hoods placed over their heads. At precisely 10:54 the trapdoor swung open under their feet and, with a snap of the rope, they died. Life settled down in Girardville following the trial and William Heath, over the next couple of years, eased himself back into the life he had known before going west.

Somewhere around 1880, for reasons unknown, William Heath moved his family from Girardville to the town of Tamaqua, which is about twenty miles due east. It could have been for a thing as simple as residing closer to where he may have been working (old mines were constantly closing up when veins were exhausted and new ones were frequently opened all around the county). It might have been something as complicated as too many people asking too many questions. Maybe Heath felt he could live in peace better in a town where he wasn't so well known. Regardless, he took up residency on Orwigsburg Street in Tamaqua, Pennsylvania. Exhibit 22 is a photograph (as it looks today) of the house William lived in after moving to Tamaqua. The house is a half of a double block and sits quietly on a narrow street; the last street on the south end of town. Exhibit 23 is a closer view of the left side of Heath's house. It is presented here because I could not help wondering how many nights William sat on this tiny porch contemplating what was and what might have been. How many people passed by him perhaps saying hello, not knowing they had just nodded to the only man to walk away from Custer's Last Stand.

Before the turn of the nineteenth century, in 1799, Berkard Moser left his home in Lehigh County, Pennsylvania, to look for a better life for himself and his family. He found his way to what is now Tamaqua. Moser arrived at the Little Schuylkill River and settled his family into temporary quarters. By 1801 he had erected a sawmill on the river as well as a traditional log home. The log home is still standing in a section of town known

Exhibit 22. Heath's home as it is today. (V. J. Genovese)

Exhibit 23. Side view of Heath home. (V. J. Genovese)

as Dutch Hill. He and his son discovered coal in 1817 and began mining it. Reports say that by 1832 they had sold fourteen thousand tons of the black rock. By the time Heath left Schuylkill County to go west, Tamaqua had given the eager public twenty-three million tons of coal to burn.

The town's forefathers laid out the streets in 1829 when the town's population did not exceed 150 people. They named it Tamaqua, which is Indian for "running water." It was running water that almost destroyed the town in 1850. The September rains came and didn't let up for days. The water flowed along the railroad tracks leading to the main mine shaft. Dragging debris with it, a huge dam was formed at the mine's entrance. As it grew in size the people failed to recognize the danger. Suddenly, with a loud roar, the dam burst. Rushing down the valley it joined with the already swollen Little Schuylkill River, sweeping away everything in its path, including a good part of the town. Sixty-two people were killed, with whole families obliterated.

The residents rebuilt and the town bounced back with the vigor of a coal-mining boomtown. Tamaqua was second in population in the county only to the county seat of Pottsville. It boasted some seven stores, five blacksmiths, four shoemakers, five doctors, and a watchmaking factory. Exhibits 24 through 27 are of Tamaqua. Exhibit 24 is the Moyer Hotel in downtown Tamaqua as it looked during Heath's residence in the town; exhibit 25 is a picture of the town as it looked circa 1880 while Heath was in residence; exhibit 26 is a photograph of the Little Schuylkill River that flows through the center of town; exhibit 27 is the town's railroad station looking very much like it did in William's time.

That William was alive and working in Tamaqua is shown by exhibit number 28. The 1888 tax records for the south ward of Tamaqua lists William Heath as a laborer with a borough tax for the year of ten dollars. Heath was going to work every day in the bowels of the earth mining coal. After toiling each day in the mine, he returned to the tiny house on Orwigsburg Street to scrub off the coal dirt from his body, have something to eat, and perhaps play with his children just like the thousands of other miners in the area. Life was changing all around him but to Heath each day was a carbon copy of the one before it. Life in the mines was hard and boring. Exhibit 29 is local artist Jack Flynn's drawing of William Heath as he looked sometime after the battle. It was done from a newspaper copy of

Exhibit 24. Moyer's Hotel. (Courtesy Don Serfass)

Exhibit 25. Tamaqua, Pennsylvania, circa 1880. (Courtesy Don Serfass)

Exhibit 26. The Little Schuylkill River, which the Indians called "Running Water," runs through Tamaqua. (V. J. Genovese)

Exhibit 27. Tamaqua Train Station as it looks today, unchanged from the time of William Heath. (V. J. Genovese)

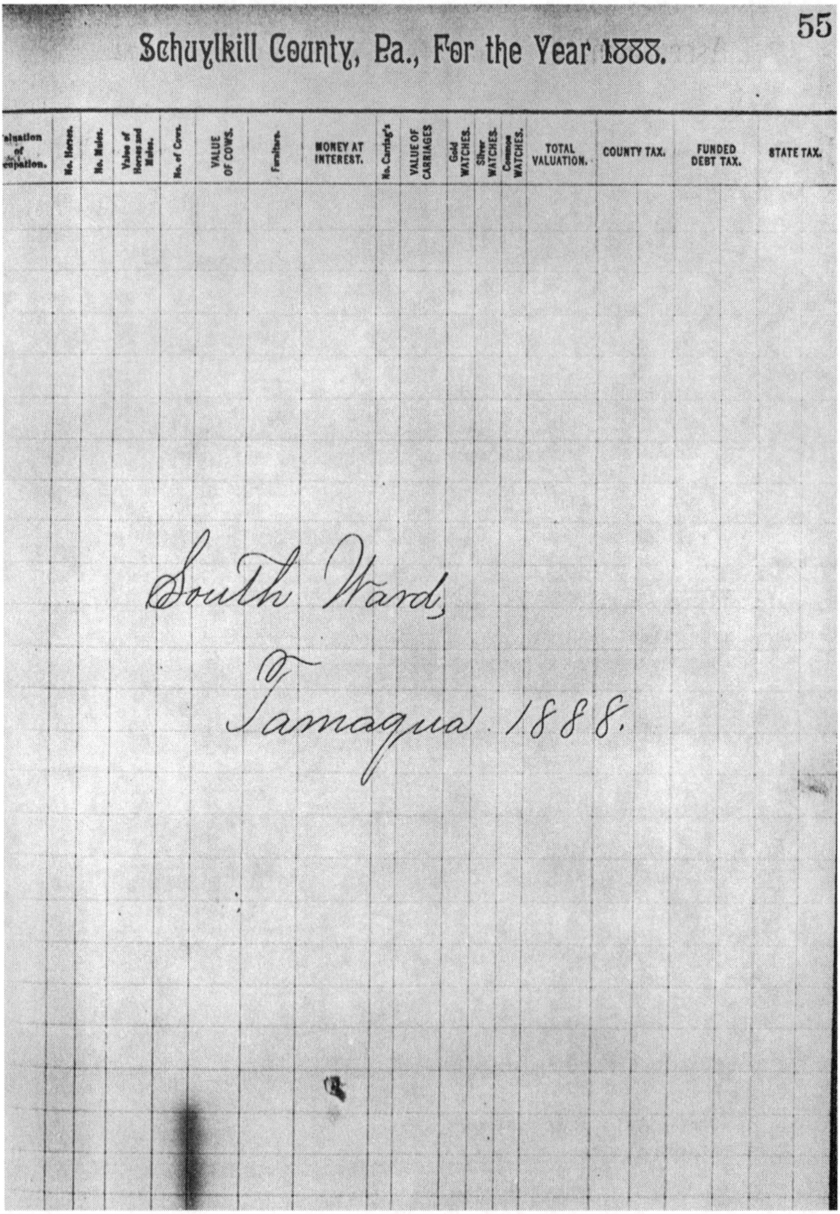

| | | | | | | | | | | | | | | | | 55 |

Schuylkill County, Pa., For the Year 1888.

aluation of ccupation.	No. Horses.	No. Mules.	Value of Horses and Mules.	No. of Cows.	VALUE OF COWS.	Furniture.	MONEY AT INTEREST.	No. Carriag's	VALUE OF CARRIAGES	Gold WATCHES.	Silver WATCHES.	Common WATCHES.	TOTAL VALUATION.	COUNTY TAX.	FUNDED DEBT TAX.	STATE TAX.

South Ward,

Tamaqua. 1888.

Exhibit 28. 1888 tax cover page (*above*) **and Tax Assessment 1888 for Schuylkill County** (*facing page*). (Courtesy Schuylkill County Courthouse Archives)

Assessment of *South Ward Tamaqua*

No. of Taxables.	NAMES OF TAXABLES.	Occupation	KIND AND DESCRIPTION OF PROPERTY.	No. of Acres	Rate Per Acre	Val Real Es
	Hampton Elias	*F. Cond*				
	Heath William	*Lab*				
	Hemshaw A.	*Rev.*				

Schuylkill County, Pa., For the Year 1888.   80

aluation of cupation.	No. Horses.	No. Mules.	Value of Horses and Mules.	No. of Cows.	VALUE OF COWS.	Furniture	MONEY AT INTEREST.	No. Carriag's	VALUE OF CARRIAGES	Gold WATCHES.	Silver WATCHES.	Common WATCHES.	TOTAL VALUATION.	COUNTY TAX.	FUNDED DEBT TAX.	STATE TAX.
100													*100*	*60*	*10*	
100													*100*	*60*	*10*	
200													*200*	*120*	*20*	

the only known photograph to exist, which is in the possession of Deborah Brumbaugh, William's great-granddaughter. As I look at his face what I wouldn't give to have one hour alone with him. There are so many questions he could answer. What a pity—no, a tragedy—that he didn't keep a journal. Exhibit 30 is a copy of the national census for the year 1880. It shows Heath alive and working in the town of Girardville before moving to Tamaqua. It also lists his family members at that time and is another government document signifying he was alive and well after the battle.

During the time period from Heath's escape from the Little Bighorn battle to his last years in Tamaqua monumental changes were taking place all around him. The most important aspect of these changes was the rising development in transportation. At the end of the Civil War in 1865 there

Exhibit 29. Artist's rendition of William Heath. (Courtesy Jack Flynn)

were about thirty-five thousand miles of railroads in the United States. Following the war and aided by the mushroomlike growth of the railroads, the westward movement began with a gusto only Americans could muster. New inventions on the rail proliferated. The Pullman sleeper made travel by rail almost luxurious. The all-steel passenger coaches and refrigerated (or reefer) cars were born. The transcontinental railroad became a reality with the driving in of the Golden Spike in 1869. The Union Pacific and Central Pacific made transportation of goods and people possible to the most remote sections of the country. With the advent of reefers freshly slaughtered beef as well as fresh fruit and vegetables could be shipped anywhere. The all-steel cars could haul huge amounts of grain, lumber, and heavy machinery.

Following the Civil War, the building of railroads heralded the era of big business. Railroad construction was expensive, so the government encouraged the public entrepreneurship with subsidies. Liberal term loans were lavished upon companies daring enough to try to join east and west by rail. The federal government also gave away over 155 million acres of land to the tracklayers. The states threw in an additional forty-nine million acres, an area much larger than the state of Texas. As the tentacles of the railroads reached out across the nation, the young Iron Colt soon became a steel horse.

Railroad companies like the Union Pacific, Northern Central, Southern Pacific, and the Santa Fe took up the challenge to span the continent. Loans were procured for the sixteen thousand dollars per mile via the flat plains to forty-eight thousand dollars for cutting through the mountains. Along the way the companies were granted the twenty square miles alongside the tracks for each mile laid. Construction gangs made up of Irish "Paddies" and Oriental "coolies" labored with sledgehammers, picks, and shovels, sometimes laying as many as ten miles of track in one day while fending off marauding Indians as they went.

Firmly attached to each railroad construction project were the famous tent towns. Counting thousands of residents these portable towns provided the laborers with all their needs; saloons, casinos, bathhouses, laundry facilities, and a number of prostitutes were easily accessible. When the crews worked their way out of sight, the town simply folded up and moved along with them. There was money to be made by all involved in the project.

Near the end of the nineteenth century no fewer than four major lines crisscrossed the country, one of the great engineering achievements of humankind. The Northern Pacific joined Duluth with Seattle. Omaha met San Francisco via the Union and Central Pacific. The Santa Fe cut through the harsh desert, wedding Atchison with Los Angeles. The south, with the completion of the Southern Pacific, had a direct route from New Orleans to Los Angeles. The Industrial Revolution could now proceed from coast to coast via the railroads.

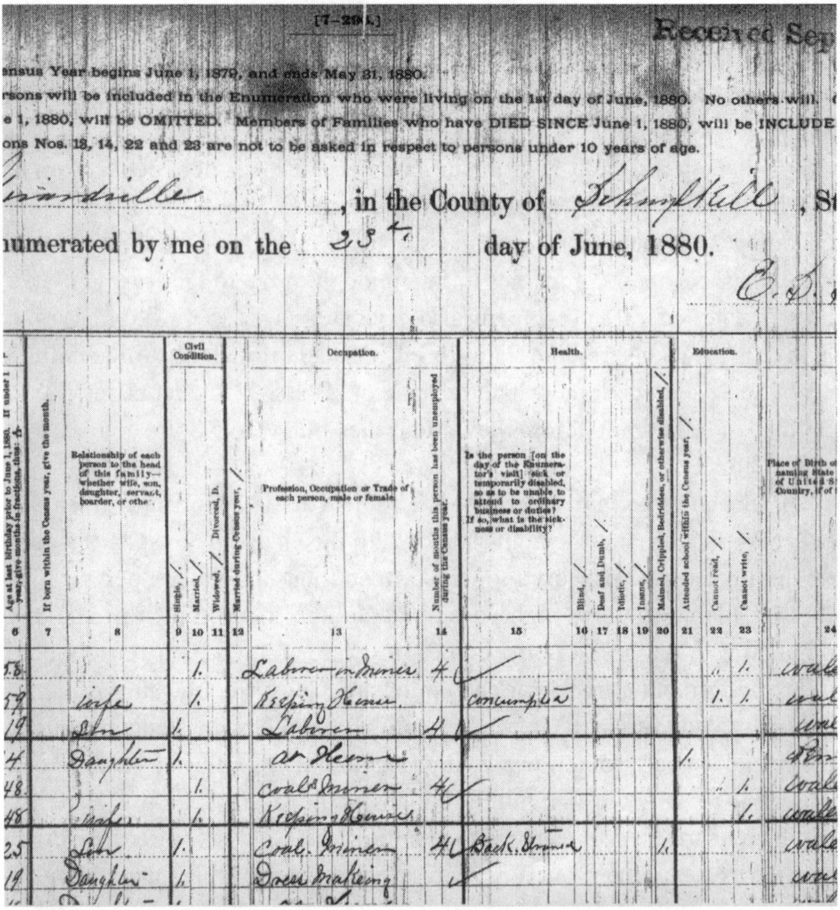

[7-296.]

Received Sep

ensus Year begins June 1, 1879, and ends May 31, 1880.

rsons will be included in the Enumeration who were living on the 1st day of June, 1880. No others will

e 1, 1880, will be OMITTED. Members of Families who have DIED SINCE June 1, 1880, will be INCLUDE

ons Nos. 13, 14, 22 and 23 are not to be asked in respect to persons under 10 years of age.

, in the County of Schuylkill , St

numerated by me on the 23d day of June, 1880.

6	7	8	9	10	11	12	13	14	15	16	17	18	19	20	21	22	23	24
58				1.			Laborer in Miner	4									1.	Wale
59		Wife		1.			Keeping House		Consumption							1.	1.	Wale
19		Son	1.				Laborer	4										Wale
4		Daughter	1.				at Home								1.			Pen
48				1.			Coal Miner	4									1.	Wale
48		Wife		1.			Keeping House										1.	Wale
25		Son	1.				Coal Miner	4	Back. Strained						1.			Wale
19		Daughter	1.				Dress Making											Wale

Exhibit 30. 1880 Census cover page (*above*) **and U.S. Census 1880** (*facing page*).

As the railroad industry boomed, so did the shipping industry. This was the heyday of the Mississippi steamboats, such as the famous *Natchez*, carrying people and products up and down the mighty river. The Great Lakes had their big cargo ships carrying many tons of coal, iron ore, and grain. Three-masted schooners by the thousands plied the ports of the coastal cities, transporting fish, coal, and lumber. America was like an adolescent boy growing so fast it was almost impossible to keep up with its needs.

The iron and steel industry became big business as well during this era. Large amounts of iron ore were shipped from Michigan through the Sault Sainte Marie Canal to the steel centers for refining. Coal was shipped by rail from Pennsylvania and West Virginia. The two met in cities like Pittsburgh, Cleveland, Toledo, and Ashtabula and were used to produce the finest steel in the world. Developing companies joined together forming huge trusts. These monopolies gained a stranglehold on American industry by using and abusing their power. Standard Oil set the tone and was soon followed by the railroads, steel, coal, and others. Names like Rockefeller, Vanderbilt, and Carnegie introduced the American age of the robber baron.

When the Industrial Revolution hit full stride the money moguls rose to the apex of their power. These titans of trust fine-tuned the art of monopoly like never before or since. The trust designers recruited stockholders in small companies in a particular commodity (like oil or steel), melding them into one giant corporation. These monopolies then had the power to control production, fix prices, and drive competitors out of business. In return the stockholders received a nice return on their share of the holdings. The country would not long tolerate the existence of the giants of industry, who lacked any social conscience, but until they were outlawed their leaders amassed obscene amounts of wealth.

Among these masters of monopoly was Cornelius "Commodore" Vanderbilt. The heavyset, loud-mouthed, white-whiskered entrepreneur had made millions in shipping and the steamboat trade. While in his late sixties he saw opportunity in the growing railroad industry and he jumped in. With vision Vanderbilt merged the New York Central Railroad and the Transcontinental lines. By giving the public lower rates with better service he made millions more than he had acquired in shipping. His fortune eclipsed the one-hundred-million-dollar mark over a century ago and all before the days of income tax. Vanderbilt University was started with a one-million-dollar grant from the commodore.

Another kingpin during the era of big trusts was Andrew Carnegie. The small Scotsman came to the United States as a boy with his poor parents in 1848 (the year Heath was born). He worked hard for years and, after accumulating some money, entered the steelmaking business in Pittsburgh. He possessed good business sense and had a natural talent for organization and administrative skills. By the turn of the twentieth century Carnegie and his partners were producing 25 percent of the country's steel. His earnings were an unheard-of twenty-five million dollars per year. J. P. Morgan, the Wall Street tycoon, eventually bought out Carnegie's holdings for $440 million in cash. Haunted by the guilt of amassing so much money, Carnegie spent the remainder of his life giving away an estimated $350 million to philanthropic causes.

The oil industry produced perhaps the granddaddy of the trust titans. John D. Rockefeller was by far the shrewdest, cruelest, and most successful of the robber barons. Not born into money, by age nineteen he had become a successful businessman. In 1870 he organized Standard Oil of Ohio. Operating out of Cleveland, his refineries squeezed out all the competition. Rockefeller would literally cut prices to the bone until his rivals went bankrupt. He hired spies to infiltrate his foes' companies and even forced the powerful railroads to pay him rebates based on the freight bills of his competitors. Andrew Carnegie referred to him as Reckafellow as he pursued his rule-or-ruin policy.

By 1882, when he formed Standard Oil, J. D. Rockefeller controlled an unbelievable 95 percent of all the oil refineries in the United States. His profits were so enormous that no one even ventured a guess at what the numbers were. He wielded more influence over the lives of this country's citizens than any president ever could. In 1890 the passage of the Sherman Anti-Trust Act signaled the beginning of the end of monopolies. However, the money these men amassed will last many lifetimes.

Strangely, one of the few things that did not drastically change during Heath's tenure at Tamaqua was agriculture. Although production increased (for example, wheat production went from about one hundred fifty million bushels at the end of the Civil War to six hundred million in 1890), the method of farming and the way of living of those who farmed stayed pretty much the same. The typical family of farmers lived in a simple wooden frame house, which was heated by a cast-iron stove in the kitchen and was lighted by kerosene lamps in every room. The kitchen, especially in the winter, was the center for all family activity. The farmer usually pumped his water by hand from a well near the house. Fewer than 10 percent of these households had any indoor plumbing.

The only recreation outside the family itself was provided by church socials. These were looked forward to by the isolated hardworking farmers. It was a welcome opportunity to dance and sing while breaking bread with neighbors. It was also virtually the sole means by which eligible young people could meet and socialize. Despite the growing industrial revolution across the nation, most people were farmers. To the tiller of the land his world was basically still a horse-centered economy. The animal served all his immediate needs. It gave him transportation, pulled his plows, provided him with sport, and was loved as a pet. Surrounding the horse was an entire industry devoted to its support. There was a large market for hay and feed. Blacksmiths were necessary for shoeing. Veterinarians were needed to keep horses healthy (many farmers were quicker to call in a doctor for an ailing horse than a sick family member). People were needed to make saddles, harnesses, wagons, carriages, and plows.

In the West, if an area had been made safe enough from Indians for several farmers to settle, there usually was a small town centrally located to meet their basic needs. A hotel would offer respite to tired, dusty travelers. The services of a barber was an added plus. A railroad siding might offer the only means of shipping grain to market. The general store could be found somewhere on the main street. A one-room schoolhouse and at least one church were a definite must. If you were very lucky the town might brag of an opera house or small theater for your pleasure.

During this "Gilded Age" labor unions began to recognize their potential power and the need to use it. With the influx of immigrants who worked dirt cheap, wages were constantly being cut. Federal laws controlling management were practically nonexistent. Labor unrest erupted in violence and was not confined to Schuylkill County. Most of the nation east of the Mississippi felt its rage. Entire cities like Chicago and Pittsburgh witnessed bloody battles in the streets. Armed militia had to be called in to restore order. The hungry and unemployed sought refuge in national labor unions like the American Federation of Labor.

As improvements were made in the workplace, the typical employee saw the length of his workday decrease. Most of the urban workers, who now toiled indoors, became interested in outdoor activities. Sports like baseball, track and field events, and football became immensely popular. Vacations were still reserved for the rich, so sports became the common person's main recreational outlet. Horse racing was a big event of the time for it could be enjoyed by both rich and poor alike. The first Kentucky Derby was run in 1872.

The Christian religious community was rocked by the new philosophies of Charles Darwin and Herbert Spencer. Darwin's theory of evolution and Spencer's principles of biology (in which he coined the phrase "survival of the fittest") brought heated debate among Christians. The Roman Catholic Church in particular was very upset by these publications.

Due to the vast concentration of wealth in the East, the major cities of that location grew by leaps and bounds, literally rebuilding themselves. Skyscrapers rose, bridges spanned rivers to carry the new horseless carriages, and populations doubled in about twenty-five years. The importance of the arts was recognized by emerging financial support for conservatories. Free public education was born. Colleges and universities began springing up like wildflowers across the nation. The world was moving at a dizzying pace around William Heath in Tamaqua, Pennsylvania.

The population west of the Mississippi River between the years 1870 to 1890 grew from seven million to sixteen million. The impetus for this growth was the Homestead Act passed by Congress in 1862. It gave up to one hundred sixty acres of land to each person who met the requirements of the law. If the land was worked and a residence established, it became the permanent property of the individual. People rushed by the thousands to take advantage of the choicest locations.

And what about the Indians while all this expansion was taking place? The pressure to gobble up the land where the Indians roamed was unstoppable; the influx of white settlers was never ending as they streamed west. If the Indians failed to recognize this before, they certainly saw it now. However, this onslaught of white settlers was beyond their ability to control. The U.S. government was relentless in its eradication of the red men. In what I like to refer to as the government's move-them-and-murder-them policy, the nation began going about its westward expansion.

The rate at which Native Americans began to disappear was astounding. In 1850 there were about one hundred thousand Indians in California; by 1860 only a reported thirty-five thousand were left. In the seven-year period ending in 1876 the U.S. government had fought over two hundred battles with the natives. We had signed and broken over three hundred peace treaties with these proud people. Even the Indian Peace Commission set up in 1867 by an act of Congress with General Terry as one of its members called the treaties a farce. Peace under these circumstances was obviously impossible. The Battle of Little Bighorn had been the climax of the Indian Wars. After that stunning victory, it was all downhill for the red man.

The government moved swiftly to clean out what resistance remained. The Crow and Blackfeet were moved off their reservations in Montana. In Colorado the Utes' land was arbitrarily confiscated. The Nez Perce were soundly defeated, as was the Apache nation in the Southwest with the subjugation of Chief Geronimo. To add insult to injury the government concocted a plan that was destined to fail. The Dawes Act of 1887 was a paternalistic attempt to civilize the American Indian by turning him into a farmer. Reservations were divided into parcels of land where, according to white Anglo-Saxon Protestant thinking, the Indian could be motivated into ambitious cultivation of the land thereby leading to wealth and prestige. No matter that the Indians were for eons hunter-gatherers and not tillers of the land. The plan ignored the fact that individual ownership of land was completely foreign to a people who thought of the land as a communal entity. Needless to say, this Indian Homestead Act was a fiasco. The Indians who foolishly participated eventually sold off the land, many of them starving in the process having abandoned their traditional way of life. Much of the land was unfit for farming in any case, and where it was arable, the government was remiss in providing the Native peoples with the wherewithal to make it productive.

In the years William Heath first began working in the mines around Girardville, gigantic herds of cattle were being driven north from Texas to the Chicago slaughterhouses. Following three main trails, they crossed over Indian lands, fattening their steers on the open pastureland as they took them to market. The Indians seldom bothered the white men on these journeys. As long as they were just passing through they were perceived as no problem. A typical cattle drive had as many as twenty-five hundred steers and the trek to market was close to fifteen hundred miles of dust and saddle sores. By the late 1880s the pastureland had been overgrazed to the point of uselessness. Ranchers began to fence in large parcels of land convenient to railroad crossings. Here they could have grain shipped in to feed their herds as well as ship the cows out to market when they were fat enough. Thus the popular and glamorous heyday of the American cowboy was over almost as quickly as it started. Gone but not forgotten, the legend of the cowboy still captures the American imagination today.

Life on the range with the American cowboy is one of the most endearing images of the Old West. The reputation and appeal of this American version of knight on horseback is universal. The origin of the name "cowboy" is uncertain. Some historians claim to trace it back to Ireland as early as 1000 C.E. The writer Jonathan Swift is said to have used the term in 1705 in his writings. By the early 1800s it became synonymous with the young men who earned a salary riding with the cattle herds.

Who were these adventuresome men toiling hard as they rode hundreds of miles fighting both the elements and the troublesome redskins during William Heath's years? They came from everywhere and represented all facets of American life. They were Civil War veterans, itinerant laborers, immigrants from abroad, and even freed black slaves. The cowboy's home was the open range; his ceiling at night the vast star-filled sky. The dangerous conditions the cowboy encountered called for him to rely on his own quick wits as well as the cooperation of his fellow cowpokes.

Although cowboys were found throughout the West, the best ones hailed from Texas. It was in Texas that their profession emerged and became legendary. The hard-riding Texan was considered the model of all cowpunchers. Rugged and lean, many were practically born in the saddle. Most were faithful employees and honest men whose chivalrous treatment of women was worthy of legend. Others were criminal vagrants who lived on cheap whiskey, aspiring only to gamble and cavort with the painted ladies of the night. Perhaps the best description of a cowboy I have come across is this one by an unidentified rancher circa 1875: "He lives hard, works hard, has few comforts and fewer necessities. He has but little if any, taste for reading. He enjoys a coarse practical joke or a smutty story; loves danger but abhors labor of the common kind; never tires riding, never wants to walk, no matter how short the distance he desires to go. He would rather fight with pistols than pray; loves tobacco, liquor and women better than any trinity. His life borders nearly upon that of an Indian. If he reads anything, it is in most cases a blood and thunder story of a sensational style. He enjoys his pipe, and relishes a practical joke on his comrades, or a corrupt tale, wherein abounds much vulgarity and animal propensity."*

The cowboy's attire was as flamboyant as the man himself. His pants were wool as was his shirt. Over the pants he wore leather chaps to protect him from briars, bushes, and rope burns. Shirtsleeves were held in place with colorful garters. Over his shirt he frequently sported a vest with pockets to hold his bag of tobacco and papers for rolling his own cigarettes. Underneath it all the cowboy wore wool longjohns and socks tucked into his boots. A scarf worn around the neck, knotted in front and hanging down the back of the neck to protect it from sun and wind, completed the ensemble. It could be drawn up over the face to keep the dust out or double as a mask when doing some dastardly deed. Leather gloves, wide-brimmed hat, and high-heeled boots (for getting a secure hold in the stirrups) put the finishing touches on the typical cowboy.

Astride his horse and with revolvers strapped on both sides of his belt, the American legend plied his trade of managing the herd. Strangely the typical cowboy did not own his own horse. The rancher usually supplied his mount and only rarely did the cowboy ride one exclusively, allowing him to grow sentimentally attached to his trusted steed. Minding the cattle herd was hard work. The cows were stubborn, stupid creatures constantly getting into difficulty. They wandered off to get lost, got stuck in bogs or barbed wire, and choked trying to eat the likes of rocks and wire fencing.

Spring and summer were spent fattening the herds on the rich grass of the plains. The birthing of new calves was also part of the spring ritual. Through these seasons the cowboy endured heat, heavy rains, branding,

---

*From *The Story of the Great American West* (Pleasantville, N.Y.: Reader's Digest Association, 1977).

fence mending, and a variety of other chores. By fall it was time for the roundup and drive to get the cattle to market. This was easily the most popular time of the year for the cowpunchers. At last they were doing what they liked the best: riding free over the wide-open country in the exhilarating trail drive. A drive could be comprised of one hundred cowboys using hundreds of horses and pushing several thousand head of cows over great distances to get to market. It is estimated that in 1871 some seven hundred thousand head of cattle were driven to market in the cowboy fashion.

The cows were notoriously spook-prone. They sometimes bolted at the slightest provocation; a rattlesnake hiss or a clap of thunder could set them off. Once a stampede started it quickly became serious business. A cowboy caught on foot in the face of it was a dead man. Even on horseback the tidal wave of animals could fast make mincemeat out of a rider. In one report of a stampede the herd literally trampled to death 341 cattle in its mad rush to who knows where. To the true cowboy it was still all worth it for at the end of the drive came payday and a couple of nights on the town in places like Dodge City, Abilene, and Wichita. Such was the life of the American cowboy.

Chapter Eight

LEGACY OF A SURVIVOR

A s the nation underwent dramatic change, the lone survivor of Little Bighorn lived on in virtual obscurity in the small town of Tamaqua, Pennsylvania. If in some magical way Heath were alive today, he would be an overnight celebrity. He would be in great demand on the talk-show circuit, and his autobiography would be a much sought-after book. He had the answers to so many of our questions regarding the flamboyant general and the infamous Last Stand. Alas, all such revelations went with him to the grave.

Since writing this story I have taken every opportunity to get people's reaction to it. The overwhelming majority are excited about the findings and look forward to reading the book. Some have asked some interesting questions. Why didn't the army come looking for him if he really was at the Little Bighorn? The answer to this one seems pretty clear to me. The army had no reason to go looking for a dead man. As far as they were concerned William Heath was dead and buried in Montana with the rest of the Seventh. Since neither William nor his wife, Margaret, made any claims for pension or disability and never spoke openly about the event, the army had no reason to come looking for him.

Name of Child	Parents Names	Where Born
Elizabeth Jane	George & Elizabeth Fry	–
Arthur	William & Margaret Heath	Mt Carmel
Mary Jane	Daniel & Emily Treloar	Tamaqua

Date of Birth	Date of Baptism	Witness Names	Minister
July 29. 1885	December 10. 1885	Mrs Walter	J. H. Acornley
May 8. 1884	December 22. 1885	Mother	J. H. Acornley
October 11. 1872	March 23. 1886	Mrs Davis & Others	

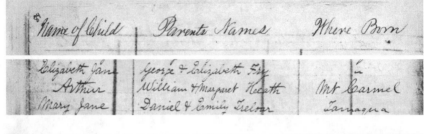

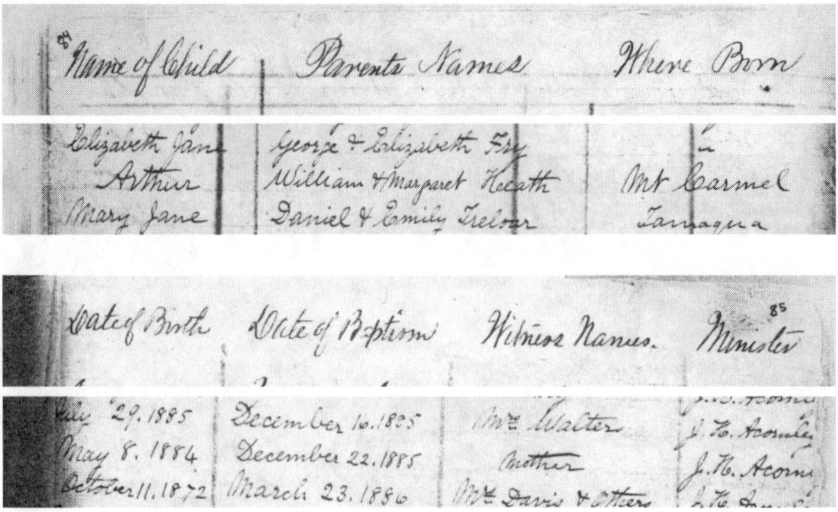

Exhibit 31. Baptismal records.
(Courtesy Tamaqua Primitive Methodist Church, Rev. Bruce Sellers)

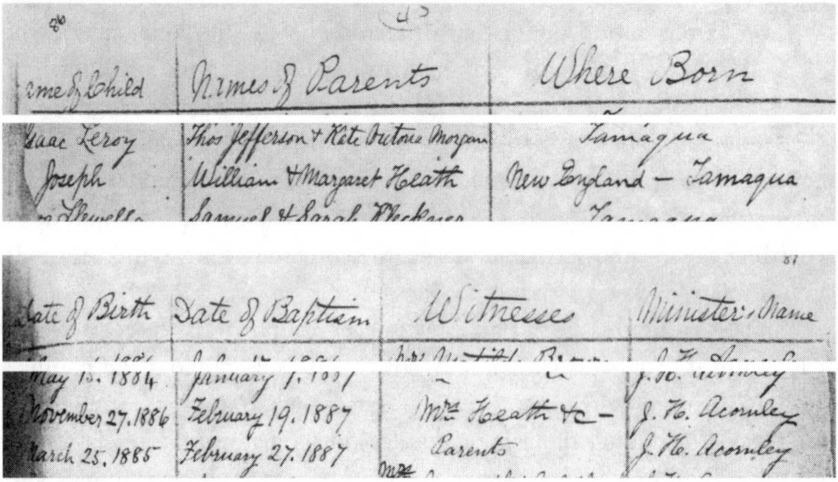

Name of Child	Names of Parents	Where Born
Isaac Leroy	Thos Jefferson & Kate Victoria Morgan	Tamaqua
Joseph	William & Margaret Heath	New England – Tamaqua

Date of Birth	Date of Baptism	Witnesses	Minister's Name
May 13. 1884	January 1. 1887	–	J. H. Acornley
November 27. 1886	February 19. 1887	Mrs Heath &c – Parents	J. H. Acornley
March 25. 1885	February 27. 1887		J. H. Acornley

Exhibit 32. Baptismal records.
(Courtesy Tamaqua Primitive Methodist Church, Rev. Bruce Sellers)

Why did the people in town keep quiet if they knew he was a deserter? It is almost certain that only family members knew where he had been. While Heath was in the army it is not even certain that Margaret knew where he was. If she did she would have refrained from telling anyone for fear the long vengeful arm of the Mollies might reach out against the

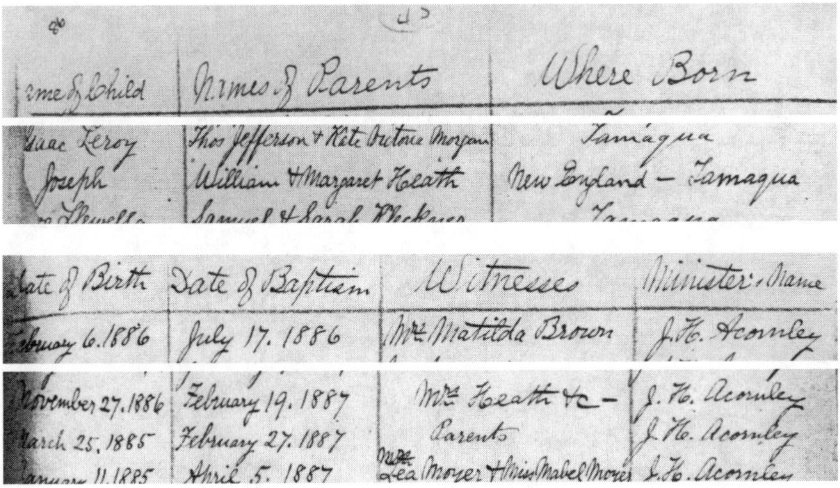

Name of Child | Names of Parents | Where Born
Isaac Leroy | Thos Jefferson & Kate Victoria Morgan | Tamaqua
Joseph | William & Margaret Heath | New England – Tamaqua

Date of Birth | Date of Baptism | Witnesses | Minister's Name
February 6. 1886 | July 17. 1886 | Mrs Matilda Brown | J. H. Acornley
November 27. 1886 | February 19. 1887 | Mrs Heath &c – | J. H. Acornley
March 25. 1885 | February 27. 1887 | Parents | J. H. Acornley

Exhibit 33. Baptismal records.
(Courtesy Tamaqua Primitive Methodist Church, Rev. Bruce Sellers)

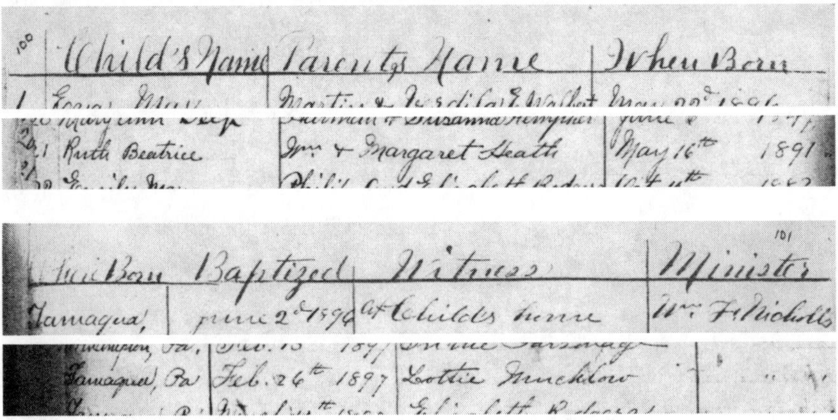

Child's Name | Parents Name | When Born
Ruth Beatrice | Wm & Margaret Heath | May 16th 1891

Where Born | Baptized | Witness | Minister
Tamaqua | June 2d 1896 | at Childs home | Wm F. Nicholls
Tamaqua, Pa | Feb. 26th 1897 | Lottie Minckler

Exhibit 34. Baptismal records.
(Courtesy Tamaqua Primitive Methodist Church, Rev. Bruce Sellers)

former Coal and Iron Policeman. Several Mollies themselves, fleeing the law, are reported to have fled west and joined the army to avoid persecution. One such fugitive is reported to have remained out west and fathered a future governor of Montana. Surely Margaret would not have taken any chances that a loose tongue would put her husband in harm's way.

Was he accepted back in the fold by the townspeople when he returned home to Girardville? Assuming they didn't know where he had been and knowing that Heath did nothing to enlighten them, it is probable

ONE PRICE ONLY, HIRSHLER & FOX, Pottsville, Pa.

924 *Tamaqua* SCHUYLKILL COUNTY *South Ward*

F G Politics See Blank Page G H

Flook Mary. Spruce st.
Flook Sarah L. Penn st.
Focht William H, 56, liveryman. Hunter st. William H, 20, stableman, Rebecca, Ellen.
Fogarty Patrick, 35, priest. W. Broad st. Jane, Esther.
Folk Samuel, 57, shoemaker. Hunter st. Mary, Lucy.
Foster Archie, 51, laborer. Spruce st. Mary, Mary A, Mezes 53, tailor, Archie 29, laborer, John 17, laborer.
Fox Christ, 26, laborer. Hunter st. Alice
Frantz Stephen, st.

Frantz, , , , ,
, , Amy L, Emma 9,
Stephens 7, Aaron D, 5, Edward M, 3.
Frehner Elmer D, 30, saddler. Penn st. Emma D, Florence D.
Fritz Annie. Spruce st. Francis L.
Fritz Edwin, 45, laborer. Spruce st. Hannah, Oliver 22, laborer, Emma.
Fry Ellsworth, 29, conductor. Penn st. Anna.
Fry Emanuel J, 64, merchant. W. Broad st. Isabelle, Edward 23, salesman.
Fry John, 57, laborer. Hunter st. Mary A.
Fry Robert, 58, laborer. Hunter st. Susanna, Willoughby 24, tinner.
Fry Joseph, 55, laborer. Hunter st. Elizabeth, Lewis 30, laborer.
Fry William C, 49, carpenter. Spruce st. Macanie, Emmeline, Courd 19, laborer, Rachael, Cora A, Anna, William 6.
Gallegher William, 40, laborer. 158 W. Broad st. Gertrude, Bertha.
Gallagher Dennis, 55, miner. Spruce st. Susan, Maggie, Ellen, Dennis 5, Gabriel 3.
Gallagher Hannah. Spruce st. James 29, miner.
Galloway William, 51, merchant. Broad st. Annie.
Ganoung Abbie. Hunter st. Cora M.
Gardner Harry C, 26, bartender. 20 W. Broad st.
Garhard William E, 33, carpenter. Hunter st. Jane A, Bertha, Mamie, Anna J, Robert 6.
Geisinger Jessie, 52, shoemaker. Hunter st. Kate, Ellen, William 25, laborer, Henry 20, laborer, Mary, Oscar 10.
Geisinger Jesse, 53, shoemaker. Penn st. Katherine, Ida.
Getkin William S, 25, conductor. Hunter st. Sarah M.
Giltden William H, 33, brakeman. Spruce st. Lynia, Sallie S.
Giltner Fracta, 26. Spruce st. Bertha, Effie, John J.
Giltner Elias, 57, carpenter. Penn st. Sophie, Jacob 22, carpenter, Chas. 20, brakeman, Joseph 17, laborer, Frank S, 11.
Green Victor, 23, salesman. W. Broad st.
Groezgil Fordnell, 27, laborer. Spruce st.
Griffiths Margaret E. 46 W. Broad st. Emma, Elmer J, 26, undertaker, Edith T.
Guldner Benj. 63, laborer. Orwigsburg st.

Caroline, Katherine, Zacharith 26, laborer.
Guldner Joel, 36, locomotive fireman. Orwigsburg st. Mary. Ann, Maud E, Frances L.
Guldner Samuel, 30, engineer. Orwigsburg st. Maggie, Laura.
Guss Wallace, 46, asst. cashier. 20 W. Broad st.
Guynan Catharine. Orwigsburg st. Mary E.
Hadesty Alfred L, 35, plumber. Bow st. Mary E, Alfred S, 18, laborer, John , , Lizard W, 13, Florence O, ,
, , , , W. Broad , , Jennie, Emelia,
Hagarty Thomas, 58, blacksmith. Bow st. Eden, Mary A, Margaret, John 24, blacksmith, Nora, James 19, blacksmith, David 16, Thomas 13.
Hagner Elmer, 26, dispatcher. Hunter st. Kate, Rosdolph 2.
Halderman Samuel, 38, brakeman. Penn st. Lyda, William 8, Samuel 5, May, Edward 3.
Halderman William, 46, engineer. Hunter st. Mary I, Mary, Annie, Caroline, William 16, messenger, Lucilla, Frederick 4, Esther, Florence.
Haldeman Catharine. Spruce st.
Haldeman George, 28, saloon keeper. Spruce st. Mary.
Haldeman Edward, 47, carpenter. Bow st. Mary, Amanda C, Frank A, 20, laborer, Ida G, Harry J, 14, laborer, Sylvester A, 11, Emma P, George L, 7, Stella R, Clara.
Hardranft Henry, 69, laborer. Hunter st. Susan A, Clara.
Harris Robert, 36, publisher. Hunter st. Sophene M, Mary A, Rachael, John T, 12, Robert M, 11.
Harrobin Thomas, 47, carpenter. Orwigsburg st. Elizabeth, William 16, Mary B, Edwin J, 12, Elizabeth L, Jennie, Robert T, 1.
Hartman Annia. 13 W. Broad st.
Hauck William, 26, painter. Orwigsburg st. Alice, Howard 5, Emma, William 1,
Haucker Chas., 38, miner. Orwigsburg st. Mary, Charles 10, William 8, Annie, Martha, Francis 2.
Hanker John, 48, miner. Penn st. Sarah, Ruth 2, 12, John M, 8, Annie E

Heatch William, 41, miner. Orwigsburg st. Margaret, John 17, laborer, William T, 15, laborer. Lavinia, Emma, Joseph 3, Editheron.
Heim Israel, 62, shoemaker. Hunter st. Ida.
Heisler William, 31, salesman. Hunter st. Mary E.
Heister Lewis J, 31, tinner. Hunter st. Clara, Clarence 9, Mamie, Howard S, 3.
Helwig George, 30, mason. Spruce st.
Hendricks William, 34, salesman. Hunter st. Susan M, John M, 7, John 73.
Hennan Michael M, 34, miner. Hunter st. Susan, Thomas 19, laborer, John 17.

Exhibit 35. A directory of the Eleventh Census of the Population of Schuylkill County, Pennsylvania, from the 1890 Census.

that he slipped back into town with no problem. William might have changed jobs to a different mine miles from the last one to avoid this problem. Anonymity and acceptance could have been the reason he moved, as has been chronicled, from Girardville to Tamaqua. Acceptance was almost assured when he chose to return to work inside the mines alongside the others. From what is known about how strong William Heath was physically and emotionally it is doubtful that fretting over the acceptance of the people around him was much of a concern to him.

Remember, as Dr. Gudelunas relates in his afterword, the world that Heath returned to in 1877 was very different from the one he left. The labor unrest that prompted his hasty retreat the year before had been diffused. The union was soundly beaten, the mines had reopened, and an uneasy temporary peace had returned to the anthracite coal mines of Schuylkill County. Most of the Molly Maguires were either in jail or dead and fear from retribution from them was a nonfactor. William Heath did not have to slink back into town with his tail between his legs. He simply had to come back home and mind his own business and this he managed to do very successfully.

There has been speculation that Heath and his wife did not get along and that is why it was possible for him to pack his things and leave so easily. That is a thought-provoking question but one that only Billy and Margaret could have answered. The readable signs seem to indicate that their marriage was fairly sound. At the time of his departure we know Margaret was pregnant as a result of their lovemaking. After the battle and when able to travel, William's first move was to return to his wife and home in Girardville. As a dead man he could have assumed another name and gone anywhere in the country he desired for no one looks for a dead man. If their relationship was that bad William could have left it behind and no one would have been the wiser. He chose not to take this course and that tells us a lot about his marriage. Lastly, after his return Billy and his wife stayed together until his death fourteen years later and over the course of that time produced seven children. Such statistics are not commonly found in a typical dysfunctional marriage.

Details of his last years are sketchy since William Heath appears to have been pretty much of a loner. He was not active in community affairs and was not a joiner of organizations. He is reported to have been a member of the local chapter of a lodge known as the Independent Order of Odd Fellows, which was a social organization for members of the Protestant faith. His return to the mines instead of his short-lived job with the Coal and Iron Police indicates he found his experience with the Molly Maguires too unpleasant to repeat. Even though the power of the Mollies was on the wane, Heath probably figured he had had enough life-threatening adventure for the rest of his days.

A picture emerges of a man who did not talk publicly of his adventure in the West, most likely for the obvious reason of never fulfilling his military obligation. He did relate some of the events to his immediate family, although he did so sparingly. If he ever gave an account of the details of the battle it has been lost over time with the death of those who knew it. Even though family members have always known the story of his survival, they have only recently spoken openly about it. In an interview with Heath's great-granddaughter, she told me she can vividly recall studying about Custer's Last Stand in high school. She said when the teacher remarked how every cavalryman was killed, leaving no survivors, she knew in her heart that was not true. She was well aware that her great-grandfather was there and did live to tell about it. With some remorse, she remembered saying nothing in class for fear of ridicule from her classmates and possible punishment from the teacher who was sure to think she was out of line.

Further proof that William Heath was not killed at the Little Bighorn battle can be found in the fact that he fathered at least seven children after the battle! The documentation of their births is irrefutable and I will discuss each child in detail. Exhibits 31 through 34 are church records from the Primitive Methodist Church in Tamaqua, Pennsylvania. Exhibit 35 is the 1890 census report. Here is additional evidence that even the U.S. government acknowledged that Heath was alive in its own official document. There is a typographical error in the spelling of Heath (Heatch), but make no mistake about who it is because there was no Heatch living on Orwigsburg Street in Tamaqua in 1890 with children by the names of Margaret, John, William, etc., at those ages. It was William Heath. The same type of error was made in the spelling of daughter Lavina's name (the census erroneously has her name as Lavinia). The 1880 census also had William's age incorrect for he would have been around thirty-two at the time. The rest of the information is correct and should give no cause for concern as this is the right William Heath.

As we already know, William Heath fathered two sons prior to leaving Girardville and joining the U.S. Army. His first son, John, born September 10, 1873, is listed in the 1890 census (exhibit 35). John started working as a breaker boy in the local mines at age nine. By age thirteen he was working underground. He did this for fifty-four years before dying in 1942 in Palmerton Hospital, a victim of the black lung disease so common to miners.

Heath's second son was William, who was born November 5, 1875, according to the 1890 census which lists him as age fifteen at the time. William worked most of his adult life as a fire boss in a coal mine in the Tamaqua area. He died in 1946 at age seventy-one.

The third child born to William Heath was unfortunate little Margaret. Named after her mother, she came into the world on March 22, 1878. After struggling for several weeks she died on April 18, 1878. Little Margaret was William's first child to be born after his supposed demise out west.

In 1879 we find the fulfillment of a promise William made to the woman who saved his life. Lavina was born July 17 of that year. Lavina went on to marry and have four children of her own, one of whom she named Margaret. She died in 1973 in her hometown of Tamaqua. She is mistakenly listed in the 1890 census as Lavinia.

Two years later the Heaths had a fifth child. Emma or Emily Heath was born November 10, 1881, and is listed in the census document. Emily had a thirty-year teaching career in the public schools of Tamaqua and died in 1953 in Chicago while visiting her daughter. Emily was child number three born to Heath after the battle.

Next in line was Arthur, born May 8, 1884. He is registered on the church records but not in the census document of 1890 because he died after living only one year. His death was a bizarre accident. He died of infection from the bite of a pet rabbit.

After Arthur came Joseph, son number four for Heath and his wife. Joseph's date of birth is November 27, 1886, as noted in the church records. His date of death could not be found. However, the unfortunate circumstances were well recorded in family history. Joe was working under a train and was crushed by it.

Daughter Editheron is shown in the 1890 census. No age is given (indicating she was quite young) nor are there any church records for her. She was probably born in either 1887 or 1888. Nothing further is known about her.

Ethel Elizabeth followed Editheron on January 30, 1889. She is listed in the church records and is not on the census document. The census questionnaire might have been filled out right before she was born or, I suggest, Ethel and Editheron are one and the same person with some kind of mix-up on the name. Ethel died in 1946 and is buried in the same cemetery as William.

The last of the children born to Margaret and William was Ruth Beatrice. She was born on May 16, 1891. She is present on the church exhibits and is absent on the 1890 census for obvious reasons. Ruth died in Tamaqua in 1963. Ruth was child number seven (or possibly eight) born to the dead man Billy Heath.

In May of 1890 William Heath may have come home after working in the mines to read the following article in the local newspaper, the *Tamaqua Courier*. "Indian Wars are a thing of the past, but there is one bad Indian, 'Big Foot' with a band of about forty braves, who is giving the Indian agent at Cheyenne some trouble. He will neither take up land nor vacate it for settlers and will probably suffer the usual fate of the Indian—to be obliged 'to move on.' After many years of such moving the Indians have been so surrounded by 'civilization' that they have no chance now to engage in a revolution of sufficient magnitude to be called a war. The best they can do is to cause a local riot."

How poignant the article was in summing up the Indian situation as it presented itself at the start of the last decade of the nineteenth century. By now the only sign of Indian violence was from isolated pockets of resistance which the army quickly put down. The once and great dominant race of Native Americans that roamed this country from shore to shore had finally been subjugated. I wonder what Heath's thoughts were if he read the newspaper item. The task he started out to do at Little Bighorn had finally been accomplished and he might have derived some satisfaction from that.

At the end of 1890 a famous incident took place at Wounded Knee, in southwest South Dakota, signaling the last gasp of concentrated defiance on the part of the American Indian. Only two weeks before the event some forty-three Indian agents surrounded Chief Sitting Bull's house on the reservation. As he was in the process of being taken into custody, several of his followers opened fire on the agents. A skirmish erupted and when it was over the great Chief Sitting Bull lay dead.

The last remains of resisting Plains Indians were holed up in the Pine Ridge area in South Dakota. They were hopelessly hemmed in by army soldiers on all sides. A small contingent managed to escape to the area now known as Badlands National Park, South Dakota, some fifty miles away. After Christmas they were given assurance by the army that their rights would be protected and their dignity as well so they returned to the Pine Ridge location. A band of Lakota warriors under the leadership of Chief Big Foot (the same chief chronicled in the *Tamaqua Courier* newspaper article) were taken into custody by none other than the Seventh Cavalry and moved under escort to Wounded Knee.

The Indians pitched their camp to protect themselves from the bitter December winds. As they looked out from their lodges they could see the bluecoats on all four sides watching them warily. A short distance away, on a ridge close by, stood four Gatling guns mounted and ready for action. These guns had the awesome capability to fire fifty rounds per minute. The Indians knew full well their destructive potential. Displayed conspicuously in the middle of the Indian camp from a high pole was a white truce banner. Both sides settled down to a restless night.

At 8:00 on the morning of December 29 commanding officer General Brooke gathered the Indians outside their lodges and ordered them to turn in their weapons. In the next instant an impetuous young brave by the name of Black Fox reached for a rifle under his blanket and fired a shot.

Gathering up the dead from the battlefield at Wounded Knee.
(Courtesy Denver Public Library, Western History Collection/North Western Photo Co./X-31464)

The waiting soldiers returned fire at point-blank range against everyone in front of them. The first volley is estimated to have killed one-half of all the Indians present. The surviving warriors, fearing certain death, charged the troopers. The squaws and their children rushed from the lodges to find out what was happening.

The four Gatling guns (called Hotchkiss guns) began firing. They shot out two-pound projectiles at almost one per second. These shells splintered the lodge poles and literally cut the enemy to pieces. Within minutes over two hundred Indians, including women carrying infants and children, were dead. Wounded Knee was and will continue to be a huge black stain in the history of how the white man has treated his red brothers. The wholesale slaughter was an obscene chapter from the book of Indian Wars that should give us all much to feel guilty about. Historians generally agree that on that gray wintry day in South Dakota the final death rattle was uttered thus ending the great Indian Wars. The West was now safe for the unbridled expansion our forefathers envisioned.

No one knows how William Heath took the news of the tumor. It was growing day by day in his brain and he was only in his early forties. The medical community at the time did not have the benefit of CAT scans, MRIs, and microsurgery. He was probably told that the tumor was terminal and he had a year at best to live. Heath appears to have been a strong man both physically and mentally, yet such bad news had to have been disheartening. Like most people in his situation his emotions must have run the gamut starting with surprise and moving on to fear, then anger, and finally resignation. This time there would be no escape. Margaret's thoughts, aside from her sympathy for her dying husband, must have included the realization that she would be left with many mouths to feed all by herself.

The treatment for a growth like William had in 1891 was basically nonexistent. Skilled successful surgery was in its infancy back then and brain surgery was a huge step behind that. The progression of such a disease is gruesome to witness. In the beginning the pain might come and go, giving some temporary relief. Over time it becomes constant and unbear-

HEATH, Albert	5 mths.	06-06-1889	90	08
HEATH, Arthur	1	12-25-1885	81	08
HEATH, Arthur	63	07-10-1912	90	08
HEATH, Edgar R.	55	04-12-1972	WH295	11
HEATH, Elizabeth		10-31-1932	90	08
HEATH, Hannah	4m,5d	12-02-1876	89	08
HEATH, Hattie D.	1879	06-27-1923	287	10
HEATH, John	1873	10-15-1942	287	10
HEATH, Margaret	66	04-16-1919	81	10
HEATH, Mattie	1878	03-26-1946	287	08
HEATH, William	44	05-02-1891	81	08
HEATH, William	1876	07-12-1946	287	10

Exhibit 36. Cemetery records. (Courtesy Tamaqua Odd Fellows Cemetery)

able, like a series of nightmarish headaches that never end. Then, little mental slipups like forgetting what day it is or not recalling a child's name start to creep in. In time the person doesn't even know his own name and becomes noncommunicative, looking at the world with that empty gaze. Motor functions begin to atrophy, necessitating that the person be confined to bed. Being bedridden causes physical deterioration. Muscles lose their tone, bedsores develop, and urinary infections from inactivity result.

Eventually, such a person becomes incontinent and requires the changing of both his dressings and the bedsheets numerous times a day. For the caregiver it is very demanding and exasperating, especially if it is a family member. Near the end the person alternately slips in and out of consciousness until finally he slips away for good. Death in such circumstances is considered a blessing. The end is a difficult thing for a loved one to witness. Better to have died quickly at Little Bighorn than to suffer this horrible fate. Even the miners' black lung disease, which claimed Heath's oldest son in 1942, offered more time and more dignity than this sad end.

On May 2, 1891 (one day after his forty-third year of life), the man who had cheated death at Custer's Last Stand succumbed. Exhibit 36 is a copy of the records from the Oddfellows Cemetery in Tamaqua, Pennsylvania, where Heath was buried. It indicates that he can be found in section 8, lot number 81. Sometime prior to his death, according to oral family history, an open letter appeared in the local newspaper inquiring as to the whereabouts of a William Heath who served out west with the Seventh Cavalry in 1876. It is reported that William became very upset over the article and did not respond to the request for information. The present family of Heath surmises that it might have been either Lavina Ennis or one of her family looking for the man whose life had become a part of theirs. This makes sense as they were the only ones outside his family who knew the details of his adventure out west. William must have made the same assumption. Still, so strong was his desire to remain anonymous and keep that part of his past hidden that he held back the urge to respond.

On the day of Heath's passing in 1891 there was no death notice in the local paper. Back then obituaries were reserved for the more well known and distinguished members of the community and even they had to pay to have them put in. William had taken great pains to avoid any such notoriety. His dying went unnoticed except by those who loved and cared for him. Curiously, while searching the newspaper for news of William's death I came across this article dated the same day he died:

It May Solve the Indian Question

Washington, May 2—The War Department is much pleased with the success it has met within the enlistment of Indians in the army. Already three troops have been formed and the department expects to have others in a short time. It is the intention of the department to utilize Indians who graduate from the Indians schools in the capacity of tailors, cooks, mechanics, blacksmiths, etc. in connection with the Indian Companies and the Indian Bureau will be urged to do all in their power to educate Indian youths to fill such positions. War Department officials and army officers think that the successful enlistment of Indians will go a great way toward solving the Indian question and making them self supporting.

On May 4, 1891, William Heath was laid to rest in the Oddfellows Cemetery. Exhibit 37 is a copy of the Primitive Methodist Church records noting Pastor Alfred Humphries officiating at the ceremony, entry 43. Heath was buried two rows back from the Soldiers Circle Monument. Dedicated to all soldiers, it is a fifty-foot-high marble monument topped off with an eagle displaying outstretched wings. It was erected in 1870 at a cost of nine thousand dollars. Near this soldiers monument the body of William Heath lies.

Exhibit 37 along with previous ones such as 28 and 31 to 34 reinforce the growing evidence that William survived the ordeal at the Horn. It would truly seem that reports of his death were greatly exaggerated. The deceased do not pay taxes, show up in not one but two national census, make love to their wives and sire numerous offspring, and get interred in the ground fourteen years after they expire. If Heath truly lost his life in Montana in 1876, none of this would be possible.

Funerals attended by J. Humphries while stationed at Tamaqua

	Mr. 17th 18??		Mrs. Y. ... R. ... March 5 1891
42	William Backus	"	April 26 1891
43	William Heath	"	May 4 1891
44	Frank Houtzberger	"	1891

Exhibit 37. Funeral records. (Courtesy Tamaqua Primitive Methodist Church, Rev. Bruce Sellers)

Exhibit 38 is a photograph of the entrance to the Oddfellows Cemetery. Exhibit 39 is a photograph of William and Margaret's headstones. Exhibit 40 is a close-up of his final resting place. Exhibit 41 is a view of Tamaqua from behind Heath's headstone, which sits on a hill overlooking the town. I wonder if he was buried wearing a scarf like the one he wore much of the time after returning home from the West, perhaps like the kind worn by the dashing General Custer and the men of the Seventh Cavalry. Two weeks after Heath's death Margaret went into labor and little Ruth was born on May 16. Both father and daughter were deprived of the pleasure of seeing each other's face. Ruth was the last testament to William's life here on earth and one of many proofs that he lived long after the battle.

Exhibit 38. Entrance to the Odd Fellows Cemetery where William Heath is buried.
(V. J. Genovese)

Exhibit 39. Headstones of Margaret and William Heath. (V. J. Genovese)

Exhibit 40. Close-up of William Heath's headstone. (V. J. Genovese)

The history books state that the only living creature to survive the Battle of Little Bighorn was Capt. Myles Keogh's horse Comanche. The claybank gelding was found with several wounds from both arrows and gunshots. Like his human counterpart, Heath, he was near death but was

Exhibit 41. Views of Tamaqua from behind William Heath's grave. (V. J. Genovese)

nursed back to health. The horse spent the remaining fifteen years of his life in the quiet solitude of his own corral at Fort Lincoln, Nebraska. Coincidentally, he died in 1891 and thus the lives of the last two survivors ended in the same year.

Historian Walter Camp's notes also include an interview with a Dennis Lynch from Cumberland, Maryland. Lynch fought with Custer in the Civil War and spent ten years out west with him chasing Indians. Lynch was assigned to Troop F which was in charge of the pack train. It was Lynch who was the first to discover the scared and wounded horse down near the river after the battle. His head drooped to the ground; the poor beast was shot five or six times. One shot went in one side of his chest and exited the other. Two wounds were evident in his neck. His front leg just above the hoof had taken a round as did the unfortunate horse's loins. Encouragingly Comanche recognized the men of the pack train and nickered when they called to him by name. Lynch himself dressed the near-dead animal's wounds with a zinc wash and led the limping and badly suffering horse onto the deck of the steamer *Far West* for the trip home.

The public's interest in Comanche, until now the generally accepted sole survivor of Little Bighorn, is exceeded only by interest in Custer himself. The following account of the life of this famous horse is taken from *Bugles, Banners, and Bonnets,* by Earnest Reedstrom.

Comanche. (Courtesy Denver Public Library, Western History Collection/D. F. Barry/B-337)

Comanche was the second horse owned by Capt. Myles W. Keogh having been purchased by him in 1868 while he served as Inspector General on Gen. Alfred Sully's staff at Fort Dodge, Kansas. Keogh's other horse was named Paddy and was used strictly for parade duty. Comanche was selected for battle no doubt due to his strength and bravery under fire. Most officers had at least two horses which they purchased at their own expense. In 1868 the army permitted for military service horses between fourteen and sixteen hands high and weighing not less than seven hundred and fifty pounds nor more than eleven hundred. The age of the horse could not be less than five years or more than eight. Comanche was five years old, stood fifteen hands high, and weighed nearly nine hundred and twenty-five pounds. The Claybank horse was mostly red in color with some yellow mixed in and had a black tail and mane.

Comanche's history of distinguishing himself in battle had long preceded Little Bighorn. In September 1868 Keogh's mount was wounded in the right hindquarter with an arrow during a fight with Comanche Indians near the Cimarron River. The company farrier found the broken-off arrow in the horse sometime later and thus Comanche was named. Two years later the horse was wounded in the right leg in yet another struggle with the Indians. This episode left him lame for nearly a month before returning to duty with his rider. Three years after this encounter, while his master tangled with some moonshiners in Kentucky, Comanche was again wounded. This time it was his right shoulder that took the hit. The battle-scarred gelding managed to avoid further serious injury until that fateful day in Montana. Following the grisly task of interring the dead, Comanche was led the fifteen miles to the steamer *Far West*. At times, along the way, the badly injured animal collapsed until he literally had to be hauled back to the boat.

For the better part of a year Comanche hung from a sling in his stall, mending his multiple wounds. Once he miraculously returned to his old self the horse was given "free run of the post and roamed at will." He often found his way to the camp's mess and was usually rewarded with a bucket of beer by some well-wisher. Military ceremonies always included Comanche with his empty saddle draped in black and boots reversed in the stirrups. Relieved of any more official duty, the horse moved with Troop I from fort to fort, finally pasturing out in Fort Riley, Kansas, where he literally led the life of riley. Fourteen years after the battle, when his permanent caretaker Gustave Korn was killed at the Battle of Wounded Knee, South Dakota, Comanche seemed to loose his desire to live and so died at age twenty-eight on November 6, 1891. He was stuffed and mounted and since then has been on display at the University of Kansas where countless numbers of the curious have mused over him. The famous warhorse had outlived William Heath by six months and four days. Although he may not have been the sole survivor of Custer's Last Stand, Comanche was most likely the last and lives on as a reminder of that day of defeat at the Battle of Little Bighorn.

Strangely, not far from Billy Heath's final resting place, barely a dozen miles from the town of Tamaqua on the side of a steep ridge rests an impressive-looking mausoleum. The granite edifice is the final resting place for American Indian Jim Thorpe. This Native American spent much of his youth running away from the white man's repressive education at the Indian school in Carlisle, Pennsylvania. His indomitable spirit and extraordinary physical abilities led him to become America's greatest athlete ever. His body was exhumed and moved from out west to Mauch Chunk, Pennsylvania, which promptly changed its name to Jim Thorpe, Pennsylvania.

The world of today has changed so much since June 25, 1876. We have walked on the moon and located the *Titanic* on the floor of the ocean. Genetic engineering seeks to redefine our very existence as a human race. The Internet has forever bound us together as one. One of the few unyielding constants through all this has been the contention in our history books that no soldiers survived Custer's Last Stand. Now that, too, is about to change.

Much of William Heath's life remains shrouded in mystery and unanswered questions which will probably never be revealed. Unfortunately, details of Heath's life after the battle are lacking. There is no diary or journal in the family's possession to fill in the blanks, and William himself left nothing in writing to help. He was not politically active, was not a member of organizations, and was not a man of means where he might have occasionally appeared in the local newspaper. If he sent any letters home from out west they no longer exist. Heath never wrote a book or gave any interviews, preferring to maintain his silence until his death. His reclusive behavior will always leave us wondering. This lack of details was a major factor in the brevity of this book and might prompt some readers to dismiss the contention that he was a survivor of Little Bighorn. The abundance of solid credible evidence I have presented indicates otherwise.

As happens with books written about Little Bighorn this book will probably raise as many questions as it answers. Stephen Ambrose commented that to study this battle is to walk into quicksand. It is the perfect analysis. Details of William Heath's escape went with him to his Tamaqua grave. I will surely go to mine always wondering what was the secret he spent the rest of his life guarding so faithfully. He claimed to have been proud of his military service yet, when that article (again according to oral family history) appeared in the newspaper looking for the Heath who served with the Seventh out west, he resisted all temptation to answer. Neither William nor his wife ever applied for a military pension (probably for the obvious reason of never fulfilling his five-year term) yet the fact that he served with the Seventh at Little Bighorn is irrefutable.

Right up to the moment Custer split his command the army had to be certain he was present. How else could they have listed him as killed in action? Once the charge on the Indian village commenced anything could have happened. The minute Custer yelled, "Hurrah, boys. We've got them," did Heath take his mount and run like hell? Although this is a possibility it did not appear to be Heath's style. Could he have been to the rear of the action tending to the horses when Company L set up a skirmish line? Did he see from his vantage point the wholesale slaughter that was taking place and decide to save himself? To some scholars I have consulted, like local historian Stu Richards, this is a more likely scenario. Might Heath have been among the last group that made a desperate charge to the river, somehow managing to stay one step ahead of club-wielding Indians and then concealing himself until darkness? Now I, too, have entered the quicksand.

While doing the research on the Battle of Little Bighorn I came across some commonly known information that grabbed my attention. Nathan Short is one of the Seventh Cavalry soldiers identified as being killed in action at the battle. Short was a private in Company C. He was born in Lehigh County, Pennsylvania, which, incidentally, borders Schuylkill. He enlisted in St. Louis on October 9, 1875, which, coincidentally, is the very same day as Heath. It is very likely that these two men found themselves traveling together as army recruits as they headed west to join the Seventh Cavalry. Finding much in common they may have become fast friends.

In early August following the battle, a dead Seventh Cavalry army mount and army carbine were found near the mouth of the Rosebud River. The site is several miles east of Little Bighorn. About one week later the body of a Seventh Cavalry soldier was discovered several more miles up the Rosebud from the previous find. It is now generally believed by students of the battle that this body was that of Pvt. Nathan Short, more evidence that someone could have escaped the scene.

What stirs my imagination is the following scenario: Heath and Short had become friends and seek each other out to converse, enjoying the company it provides after a long day's march. Some time during the battle they either found themselves thrown together or perhaps made sure they joined up. At some point it becomes obvious to them that if they wanted to live they must make a run for it. Together they take off on horseback, charging through a small rift in the Indians' line. Their flight does not go unnoticed and the enemy rises after them in hot pursuit. Somehow, despite the odds, they manage to get some miles from the battle before the Indians catch up. By now there may be only a few keeping up the chase as others may have dropped off and turned back. A short skirmish takes place and one of the two men loses his rifle as his horse is shot out from under him. They drive the Indians off and, doubling up on the lone horse, make a second run for freedom. Short is mortally wounded and falls dead from the horse where he is found some six weeks later. William Heath bids a sad and hasty good-bye to his old friend and moves on. There is no evidence to support this series of events, still I can't help but wonder if we came within a hair of having two Little Bighorn survivors.

Whatever happened, it is fair to say that Billy Heath felt guilty about it. He spent the rest of his life keeping the details to himself. His apparent guilt points to some potential wrongdoing, at least in his mind. The shame he carried could have been because he deserted, if not during the battle itself, then afterward when he failed to rejoin his unit. Desertion carries a harsh sentence and we should be careful when rushing to judgment a man who is not here to defend himself. There were many Indian accounts of soldiers trying to flee at different points in the battle. Trying to retreat to a safer position coupled with the urge to survive death does not constitute desertion. Cries of desertion by Custer scholars should be viewed warily as a last-gasp effort to explain away what they do not want to consider. There is absolutely no evidence that Heath deserted although the possibility does exist. On the other hand, remember, Custer brutally punished known deserters (sometimes with death) yet he himself was guilty of deserting his post in order to be with his beloved Libbie. After the battle, the failure of Heath to rejoin his unit could certainly be viewed as abandonment of his sworn duty. Was the guilt of being the only one to survive out of several hundred soldiers too much to bear? The more I wonder the deeper I sink into the quicksand.

William Heath's story is common knowledge in Tamaqua. The story first turned up in the local newspaper, the *Times News*, on the 123rd anniversary of the battle. Award-winning journalist Don Serfass broke the news with information from local historians Mark Major and Stu Richards that had been provided by Heath's great-grandchildren Deb Brumbaugh and Rich Taylor. The locals are well aware of the story and tend to view it nonchalantly. There is a crude sign (exhibit 42) weathered with age pointing the way to his grave to satisfy the curiosity seeker. I suspect the erection of a new one will soon be in order as the Custer scholars and history buffs beat a path to see for themselves the final resting place of the Lazarus of Little Bighorn.

Exhibit 42. Marker pointing the way to William Heath's final resting place. (William Gaydos)

This book will, like a thunderhead passing over the hills of Little Bighorn on a hot June day, arouse both skeptics and critics. I welcome the opportunity to do battle with them. The skeptics will say, "But that was so long ago. Why dredge it up now? Best to just leave it be as it always was. It's almost un-American to think of it otherwise." Also, there's no iron-clad evidence. The most vocal critics will be the Custer scholars and historians, learned men with far more impressive credentials than I. They will say it simply isn't so. Their rebuttals will include statements like "We've heard claims like this before and they just don't stand up" and "Yes, the Indian accounts have always allowed for this possibility but they are unreliable witnesses not to be completely believed."

I caution the reader to be wary of the motives of both groups of critics. Skeptics live their entire lives with closed minds. Scholars will view this book as a personal attack on their credibility and what has been written for the past 127 years. That is not my intent.

In the past virtually all historians and students of the battle have been wedded to the idea that no one survived Custer's last stand. In retrospect it is understandable that they took this position. The past purveyors of sur-vival claims were strapped with scant evidence to back up their claim. The popular joke of there being more survivors than participants is all part of the myth. Upon close scrutiny all these past claims were found to have no merit and were justifiably dismissed. The Heath story is not found to be wanting in evidence to back it up. There is a plethora of proof to the con-trary and as such deserves a reexamination of the long-held belief that no one escaped the Little Bighorn.

The standard criticism of Heath's story is twofold. How do we know that the Heath of Girardville and the Heath in Montana are the same man? The answer lies in setting your prejudices aside and looking at the facts. The documents presented along with corroborating family history support the contention that they were one and the same. Both men shared too many common factors to come to any other conclusion. Both had the same name, birthplace, age, height, weight, former occupation, and so forth. All the dates in Heath's life fit, his reason for leaving Girardville makes sense, and the fact that he returned very much alive afterward is unassailable. In terms of hard evidence there is ample proof to conclude the Heath in Montana came from Girardville. There is no evidence to the contrary.

Shoot! This guy was a deserter is the second major criticism. Often it is levied after the first criticism fails to stand the test of proof. One should be careful about throwing out such a statement. The charge of desertion is serious and very damaging to the person in question. It is like the mark of leprosy that never disappears. Once leveled it seldom goes away even if later proven to be false. It is wrong to arbitrarily tarnish a man's reputation plus there are members of Heath's family still alive.

I have admitted that once William was well enough to rejoin his unit and did not do so he became, by definition, a deserter. However, prior to and during the fight there is not one shred of evidence pointing to desertion. Keep in mind it is the U.S. Army that said Heath was there and believed he died in battle. They certainly would not maintain such a position if they knew or even thought he had deserted prior to the battle. Admittedly there are many things we may never know about Billy Heath's life. His survival of Little Bighorn is not one of them.

Questions will be raised about my qualifications to make such a bold contention. Here, I must admit, I have to plead ignorance. I am not aware that there are rigid qualifications for discovering the truth. I am mindful of the fact that to practice medicine one needs to meet certain standards. I know that to teach a classroom of children one has to have met licensing criteria. Even someone operating heavy equipment has to have certain competencies. The truth is what the majority of people accept as true and I trust that the final verdict of the readers will bear this out.

I am certain that to tell the truth one simply has to be reasonably intelligent, possess an earnest desire to know the facts, and be psychologically and morally sound.

To my critics I say check the facts as I have offered them and you will conclude that it happened; at least one man *did* get away. An honest mistake was made 127 years ago. There were no dog tags. Bodies were mutilated and sitting out in the hot sun for days, emanating an unbearable stench. Animals and birds were probably feasting on the remains of the Seventh Cavalry. In the understandable haste to get the soldiers properly buried, someone made a blunder. With no disrespect intended, many of the men undertaking this task probably could not even count to 260. It's time to set the record straight and move on.

The implications of Heath's story are considerable. A belief held for 127 years does not (nor should it) die easily. History books have with sound reasons, until now, reported the confirmed death of every man in the five companies under Custer's command at Little Bighorn. The veracity of that belief I propose is now in doubt. If Heath's story is creditworthy then the history of Custer's Last Stand needs to be redefined with a footnote on William Heath.

An additional inference from all this arises. If William Heath did survive could there have been others? There were a number of troopers from the ill-fated Seventh who were never found and identified. All were assumed dead and that is probably the case. Nevertheless, in light of the Heath revelation one can't help but ask the question could there be more. Dr. Brian Dippie, author of *Custer's Last Stand: Anatomy of an American Myth*, may have uttered the ultimate paragon of paradoxes about the Horn when he said to the effect that because we've never been able to find any survivors we can never really be certain that there were no survivors. I would add that since we may have found a survivor, can we ever be sure there are no more?

You may justifiably inquire about my motive for bringing forth this revelation. It is simply that the history we teach our children should be the truth. And what exactly is the truth? I firmly believe the following statements to be truthful and I am just as firmly convinced that the evidence supports me:

1. William Heath was a living, breathing person who was born in 1848 and died in 1891.
2. William Heath was a member of the U.S. Army's Seventh Cavalry in the summer of 1876.
3. William Heath made muster, mounted up, and rode off to battle with General Custer on June 25, 1876.
4. William Heath, willingly or otherwise, took part in the Battle of Little Bighorn on that day and lived to tell about it.
5. William Heath returned to Girardville, Pennsylvania, in the spring of 1877 and over the next fourteen years worked, paid taxes, fathered seven children, and lived an uneventful life until his death on May 2, 1891.

Conclusion: William Heath was the only known military survivor of the Battle of Little Bighorn.

AFTERWORD

Those who believe that no soldier survived the Battle of Little Bighorn will certainly question why Billy Heath would have returned to Schuylkill County after he fled in the face of death threats just two years earlier. Under ordinary circumstances, the question would be quite appropriate and offer strong circumstantial evidence that the Billy Heath who "reentered" the county in 1877 was not the same person who hastily departed it in 1875. Perhaps a relative "returned" or someone assumed his identity, but the "real" Billy Heath perished on the field of battle with his comrades from the Seventh Cavalry.

The "real" Billy Heath argument quite naturally assumes that Schuylkill County changed little between 1875 and 1877. However, this assumption is not valid. Unprecedented political, social, and economic transformations had occurred (or reached fruition) during these pivotal years. Most importantly, the power of the Molly Maguires had been shattered, and a private police force, the Coal and Iron Police, had brought a form of law and order to a previously socially traumatized region. These changes were largely associated with the rise to complete economic dominance of the Philadelphia Coal and Iron Company and its president, Franklin B. Gowen. Heath, who fled the Mollies and this lawlessness, would have returned to a dramatically altered situation.

Franklin B. Gowen, once District Attorney of Schuylkill County, had risen to the presidency of the Reading Company which included its subsidiary, the Philadelphia and Reading Coal and Iron Company. In essence, Gowen believed that the anthracite coal industry was seriously restricted by ruinous competition. In the early 1870s as many as one hundred mining operations produced coal in the Schuylkill region. The Reading Company made a decision to buy thousands of acres of coal lands and either buy out or force out most individual operators. Fewer than forty collieries remained independently owned by 1875.

In order to assure the success of this endeavor to create a "cartel in anthracite coal," Gowen waged a campaign to eliminate power wielded by the area's only viable union, the Workingmen's Benevolent Association (WBA) and the secretive Molly Maguires. Gowen adroitly destroyed the WBA essentially by starving miners into submission in the famed "Long Strike" of January to June of 1875. The WBA lost virtually all credibility with miners after this forced capitulation. The destruction of the WBA caused an upsurge of labor violence attributed to the Molly Maguires.

Gowen immediately launched a drive to destroy the Mollies. He already had one weapon at his disposal. Beginning in 1867, the Pennsylvania state legislature had allowed a new police force to be established in the lower anthracite area to supplement the largely ineffectual local constables. The result was the creation of the Reading Company's privately funded Coal and Iron Police. By the 1870s these company-hired "guards" were given full police powers. In fact, the commonwealth had delegated policing powers to a private company. The Coal and Iron Police were the unquestioned rulers of the streets of Schuylkill County by 1875.

Gowen then hired the Pinkerton Detective Agency to collaborate with the Coal and Iron Police in the destruction of the Molly Maguires. The Pinkertons infiltrated the Molly organization through the work of spies, mainly James McParlan who lived in the county from 1873 to 1876. Eventually, twenty Mollies were executed between June of 1877 and October of 1879. Nine of these executions were carried out in Schuylkill County. All of the county's executions involved men arrested by the Coal and Iron

Police and convicted mainly on evidence provided by "witnesses" such as McParlan and other Pinkertons. Gowen actually was appointed special prosecutor by the county in several of the cases—all of which resulted in convictions and death sentences. In essence, Heath left a county in social turmoil in which the WBA and Molly Maguires wielded great power and returned to one in which the Reading Company and its police ruled—economically, legally, socially, and politically.

It can then be reasonably argued that a man with strong ties to his family could have recognized that his semivoluntary "exile" from Schuylkill County could now safely end. Heath's basic occupational skills involved mining and the burgeoning coal trade further encouraged his return. The most rapid development of the region's coalfields was also occurring in the northern part of the county, the very locale of Heath's pre-1875 "home."

All of these basic facts concerning the dramatic changes in the Schuylkill County of the 1870s certainly add plausibility to the argument that Comanche was not the lone Seventh Cavalry survivor of the Battle of the Little Bighorn.

Dr. William Gudelunas
Professor, Political Science and American History
College of the Desert
Palm Desert, California

BIBLIOGRAPHY

Ambrose, Stephen. *Crazy Horse and Custer*. Garden City, N.J.: Doubleday and Co., 1975.

American Folklore and Legend. Pleasantville, N.Y.: Reader's Digest Association, 1978.

Bailey, T. A. *The American Pageant*. Boston: D. C. Heath and Co., 1961.

Barnett, Louise. *Touched By Fire*. New York: H. Holt, 1996.

Bourne, H. E., and E. J. Benton. *A History of the United States*. New York: Heath, 1933.

Brown, Dee. *Bury My Heart at Wounded Knee*. New York: H. Holt Co., 1971.

Connell, Evan S. *Son of the Morning Star*. San Francisco, Calif.: North Point Press, 1984.

Davison and Neale. *Abnormal Psychology*. New York: John Wiley and Sons, 1994.

Deloria, Vine. *Custer Died for Your Sins*. Norman: University of Oklahoma Press, 1988.

Dewees, F. P. *The Molly Maguires*. New York: Burt Franklin, 1877.

Dippie, Brian. *Custer's Last Stand: Anatomy of an American Myth*. Lincoln: University of Nebraska Press, 2000.

Epple, Jesse. *Custer's Battle of the Washita*. New York: Exposition Press, 1970.

Fox, Richard. *Archeology History and Custer's Last Battle*. Norman: University of Oklahoma Press, 1993.

Goble, Paul. *Red Hawk's Account of Custer's Last Stand*. New York: Pantheon Books, 1969.

Graham, W. A. "The Story of Custer's Last Message," Taken from the *U.S. Cavalry Journal* as originally published. *Century Magazine* (1892).

———. *The Story of Little Big Horn.* Mechanicsburg, Pa.: Stackpole Books, 1994.

Gruver, Rebecca Brooks. *An American History.* New York: Appleton-Century Crofts, 1972.

Hammer, Kenneth, ed. *Custer in '76: Walter Camp's Notes on the Custer Fight.* Provo, Utah: Brigham Young University Press, 1976.

———. *Men with Custer.* Hardin: Custer Battlefield Museum, 1995.

History of Schuylkill County. New York: W. W. Munsell and Co., 1881.

Hoffling, Charles K., M.D. *Custer and the Little Big Horn: A Psychobiographical Inquiry.* Detroit.: Wayne State University Press, 1981.

Holton, James L. *The Reading Railroad: History of a Coal Age Empire,* vol. 1. Laury's Station, Pa.: Garrigues House, 1989.

Jacob, Kathryn Allamong. "Vinnie Ream." *Smithsonian* (August 2000): 104–15.

Kenney, Kevin. *Making Sense of the Molly Maguires.* New York: Oxford University Press, 1997.

Michno, Greg. *Lakota Noon.* Missoula, Mont.: Mountain Press Pub. Co., 1997.

Miller, Donald L., and Richard E. Sharpless. *The Kingdom of Coal.* Philadelphia: University of Pennsylvania Press, 1985.

Morrison, S. E. *Oxford History of the American People.* New York: Oxford University Press, 1966.

Noyes, Herman M. *The Story of America.* New York: Holt, Rinehart, and Winston, 1964.

Panzeri, Peter. *Little Big Horn 1876: Custer's Last Stand.* London: Reed International Books, 1995.

Reedstrom, Earnest. *Bugles, Banners, and War Bonnets.* Caldwell, Idaho: Caxton, 1977.

Ritchie, Donald A. *The Modern Era Since 1865.* New York: McGraw Hill, 1997.

Robinson, Charles M. *A Good Year to Die: The Story of the Great Sioux War.* New York: Random House, 1995.

Sandoz, Mari. *The Battle of Little Big Horn.* Pennsylvania: Lippincott, 1966.

Stanton, Lorraine. "The Girardville Chronicles."

The Story of the Great American West. Pleasantville, NY,: Readers Digest Assoc., 1977.

200 Years. Washington, D.C.: U.S. News and World Report, 1975.

Utley, Robert. *Cavalier in Buckskin.* Norman: University of Oklahoma Press, 1988.

———. "Custer: Hero or Butcher?" *American History Illustrated* (February 1971): 4–9, 43–48.

Welty, Raymond. "The Daily Life of the Frontier Soldier." *U.S. Cavalry Journal* 26 (1927).

Wissler, Clark. *Indians of the United States.* Garden City, N.J.: Doubleday Co., 1966.

INDEX

229